# Offshore

*By the same author*

The Teatime Islands

# Offshore

*In search of an island of my own*

BEN FOGLE

MICHAEL JOSEPH
*an imprint of*
PENGUIN BOOKS

MICHAEL JOSEPH
Published by the Penguin Group
Penguin Books Ltd, 80 Strand, London WC2R ORL, England
Penguin Group (USA) Inc., 375 Hudson Street, New York, New York 10014, USA
Penguin Group (Canada), 90 Eglinton Avenue East, Suite 700, Toronto, Ontario, Canada M4P 2Y3
(a division of Pearson Penguin Canada Inc.)
Penguin Ireland, 25 St Stephen's Green, Dublin 2, Ireland
(a division of Penguin Books Ltd)
Penguin Group (Australia), 250 Camberwell Road,
Camberwell, Victoria 3124, Australia (a division of Pearson Australia Group Pty Ltd)
Penguin Books India Pvt Ltd, 11 Community Centre,
Panchsheel Park, New Delhi – 110 017, India
Penguin Group (NZ), cnr Airborne and Rosedale Roads, Albany,
Auckland 1310, New Zealand (a division of Pearson New Zealand Ltd)
Penguin Books (South Africa) (Pty) Ltd, 24 Sturdee Avenue,
Rosebank, Johannesburg 2196, South Africa
Penguin Books Ltd, Registered Offices: 80 Strand, London WC2R ORL, England
www.penguin.com

First published 2006

1

Copyright © Ben Fogle, 2006

The moral right of the author has been asserted

Grateful acknowledgement is made for permission to quote from 'Rock on Rockall', composer Brian
Warfield, publisher Skin Music; and from 'Rockall', reproduced by kind permission of the Flanders and
Swann Estates.

Set in 13.5/16 pt Monotype Garamond
Typeset by Rowland Phototypesetting Ltd, Bury St Edmunds, Suffolk
Printed in Great Britain by Clays Ltd, St Ives plc

A CIP catalogue record for this book is available from the British Library

ISBN-13: 978-0-718-14778-5
ISBN-10: 0-718-14778-2

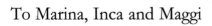

To Marina, Inca and Maggi

# CONTENTS

# Acknowledgements

Thanks to Jamie Robinson and the crew of the *Eda Frandsen* for battling on; the THV *Patricia* and her crew – sorry about the helicopter; the RAF coastguard for rescuing me – sorry, Mum, for the panic; Musto clothing; the *Orca* and her crew; Patrick and Gwyneth Murphy; Simon Glynn; Princes Michael, James and Liam of Sealand; the *Deu Kerens* and her crew; the late Babs Atkins; the Seigneur of Sark; Norman and John Gillies; Daphne Caine at the Isle of Man tourism department; Teresa Bogan and the BBC *Countryfile* team; Niall Edworthy; Julian Alexander and the team at LAW; Hilary Murray and the team at Arlington; Alison Griffin at Lake-Smith Griffin; Rowland White for once again trusting in me; Georgina Atsiaris, Jane Opoku and the team at Michael Joseph; Lady Mary von Westenholz for a little bit of paradise and inspiration at the Penthouse in Putsborough. A big thank you to all the people who helped me along the way, and to my family. And special thanks to my beloved Marina for being my translator, putting up with my frequent absences and looking after my dog.

# Introduction

The trawler pitched and reared and yawed and groaned with each enormous wave. The noise was deafening as we were tossed around like a balsa wood toy in rapids. I lay terrified in my bunk, wincing as each giant wall of water crashed against the bow.

Up and down, up and down we went as wave after wave sent liquid mountains flooding down the deck above my head. My white-knuckled hands were clamped to the mattress to keep me from catapulting from my bunk. My forehead was clammy with sweat and my stomach contorted with each jarring movement. The airless room was almost suffocating.

Unable to sleep in this watery hell, I mustered the courage to clamber from my tiny pit of a bunk and zigzagged with each violent lurch down the darkened corridor and up into the wheelhouse. As the wind howled through the lanyards the noise was thunderous. It was jet black outside and impossible to stand as we beat on into the relentless storm, rising and falling with each ferocious wave. A small pool of light illuminated the chart as the skipper in his dripping wet orange foulies peered into the inky green radar.

'It's not looking good!' he bellowed above the din of the storm, 'and the forecast is that it will get worse!

Storm twelve!' This was a Scots trawlerman, the veteran of a thousand Atlantic storms, telling me that things weren't looking good. This was a man who made a living out of bad weather. Truly, we were in some serious trouble, I thought. 'We've still got a hundred miles to run,' he continued. 'There's a severe weather warning – it'll take us a week with this headwind!'

As he finished speaking, an enormous jolting thud convulsed the boat and a vast wave enveloped us. For a terrifying moment as the engine wheezed and screamed with effort it felt as if we had come to a halt. There was a whoosh, a crash and then the sound of shattering as a wave smashed its way through the wheelhouse window and let in the howling wind.

We were in the eye of a savage Atlantic winter storm, on our way to one of the last disputed places in Europe, a land claimed by many nations but inhabited by none, an island so remote and inhospitable that more people have set foot on the moon than on its rocky shores. I was on my way to take possession of this uninhabited outcrop. The weather, though, had beaten us this time, as it had so many before us, and the boat could take no more battering. Soaked to the bone, the skipper turned her about and we limped back, storm-damaged, to harbour. My plans had been scuppered; a year's work had been lost, literally, to the wind. But I would be back. That elusive island, which had tormented and haunted me for so long, would still be mine.

*

I have always wanted an island that I could call my own. I'm not quite sure where my obsession began, or indeed why, but what I do know is that my fixation with small isles has waxed rather than waned through the years, especially with those that lie within a gull's flight of our own shores. Perhaps it stemmed from so many happy memories of childhood holidays spent in my father's native Canada, which had acquired a momentum of their own and simply rolled their way into my adulthood.

Each summer we used to head into the Canadian wilderness, to the shores of one of the country's three million lakes where my grandfather had built a log cabin. Our lake had a number of little islands scattered across it and every day I used to row out to one of them on a little wooden raft I had built, equipped with a box of provisions, a telescope and my own personal flag. After beaching my craft and striding ashore like Captain Cook himself, I would drive the flag into the ground to leave no one in my family in any doubt that this was my land and they dared step foot upon it at their peril. I used to make myself a den and then spend the rest of the day exploring 'Benland', occasionally stopping to scan the horizon with my telescope, trying to pick out imaginary invaders of my sovereign territory.

Perhaps life would have panned out rather differently had I spent those holidays climbing mountains instead, or canoeing down rivers or sitting on tropical beaches or tramping around the great capitals of the world. I might have been seeking an idealized version of those worlds today, and in all probability my waking thoughts would not have been dominated, as they are, by images of water lapping against sandy or rocky shores. Nor would my mailbox have been stuffed with estate agents' particulars offering me the chance to acquire assorted craggy outcrops around Britain. Certainly, it is highly unlikely that I would have found myself, as I do every now and then, walking out of my London flat in my windcheater and my deeply unfashionable but beloved Guernsey jumper on the way to risk my neck in a force twelve in the pursuit of some indeterminate offshore dream.

I suspect, though, that to understand my passion for islands I should plumb greater psychological depths than mere happy childhood memories. Perhaps I love the idea of being able to escape from time to time into a distant realm, far from overcrowded cities and busy work schedules, safe in the knowledge that a choppy sea stands between me and the outside world from which I seek to take brief refuge. Perhaps it is solitude that I hanker after, though that theory makes sense only up to a point, because one of the great attractions of islands for me is community: a handful of people huddled together cheek by jowl in a self-contained world where life has been stripped down to its bare essentials.

On an island, life seems much simpler than it is on the mainland, more fathomable, so much more *real*, for the simple reason that you are forced to make do with a more limited environment and the small amount of fellow human beings you have for company. On the mainland we have become gluttons, spoilt for choice in every area of our lives, but on an island you have to get on with what you have, and you become deeply grateful for the day's humble offerings. You take nothing for granted in the way that you are prone to do in a large city or town. Simple moments like coming home to a blazing fire at the end of a day's work or a blustery clifftop walk, or cooking a piece of freshly landed fish with a handful of vegetables lifted from your own small plot, become activities loaded with great significance and satisfaction when you are living on an island.

Friendships, too, become more precious, and therefore stronger, because the stakes are that much higher. In London you can fall out with someone, for instance, and the blow is slightly cushioned by the knowledge at the back of your mind that there are ten million or so other people, within a walk or a tube journey, who might step into the breach where the old friend or acquaintance once stood. On a small island you do your best, by sacrifice, compromise and greater effort, to ensure that relationships work. You have to get over yourself in order to get along. You are tested and challenged more as a human being. Deep inside all of us, perhaps, is a yearning to live in the type of harmonious community as depicted in children's books and television series like *Balamory*, *Postman Pat* or *The Waltons*. There is, of course, no such thing as a perfect community, but knowing that does not extinguish the wish to be part of one.

The year I spent as part of a castaway community on the Scottish isle of Taransay was more of a symptom than a cause of my fixation with islands. It proved to be the confirmation of a long-held suspicion that part of me yearns to be an 'islander' and, far from satisfying my curiosity, it served only to intensify my dream of one day acquiring my very own plot of land in a storm-tossed sea. The journey I took around the globe not long after my sojourn on Taransay, visiting the remote outposts of the British Empire, fired my curiosity for island life still further.

Nor can this strange calling to islands simply be reduced to a simple love for nature and/or the sea, because longings for both of these can be wholly satisfied here on mainland Britain, where we are spoiled for choice in the beauty and variety of our landscape and over eight thousand miles of coastline.

Perhaps, then, I have to face up to the unsettling possibility that I am just a megalomaniac who has taken the notion of turning my English (or British) home into my castle to a crazy extreme. The first signs of this condition emerged when I shared a bedroom with my sister as a child: I used to rope off half of it and declare it out of bounds to all outsiders. Psychiatrists may well tell me, if I were to lie on their couches, that deep down I am a frustrated dictator who will not be satisfied until I am the happy, all-powerful ruler over all that I survey, the lord of my own personal fiefdom, and can stand atop its highest point and announce grandly, 'Mine, all mine!' No doubt in my dotage you will find me stomping around my land with a shotgun tucked under my arm shouting 'Get off my land!' at passing ramblers and trespassers.

Our islands are steeped in history, myths and legends that have been passed down the generations in song, poem, storybook and the spoken word. There is not one island that is without its fabulous tales. Some are based on hard historical fact, others on outrageous fiction; some are tragic, others comic or absurd. As I grew a little older I found myself drawn to such stories like a ship to a port, immersing myself at bedtime in books about goblins, sea monsters, fairies, pirates, smugglers, shipwrecks and ghosts.

One of my favourites is the true story of the 8,000-ton SS *Politician*, which ran aground in 1941 in heavy weather off the islands of Eriskay and Barra in the Outer Hebrides, carrying a cargo of 250,000 bottles of whisky, just two days after leaving Liverpool en route for Jamaica. The story was immortalized in Compton Mackenzie's 1947 novel *Whisky Galore*, later made into the classic comedy of that name. The plot of a hapless English Home Guard captain trying to prevent the booty from falling into the hands of the locals was tailormade for an old-fashioned farce, and it would probably have been dismissed as utterly ridiculous had the events been invented rather than merely embellished by the author. Twenty Barra men were arrested and sent to prison for 'stealing' the untaxed contraband, but most of the whisky was never discovered, having been secreted under floorboards and down rabbit holes, and sunk in tidal pools.

The book and film had captivated and amused me as a child, and it was a great thrill for me when a few years ago I went to visit the islands and met a number of the now elderly heroes and characters of the story. During my stay I was shown some of the places where the locals had stored the cases of whisky and prevented it from falling into the hands of the authorities. I even got to hold an original bottle of 'Polly' whisky. It's undrinkable these days, I was told, but no doubt it will be cherished for generations to come as one of the more curious and comic artefacts of our islands' history.

In 1988 the AM Politician lounge bar, a tiny modern bungalow and the only pub on Eriskay, opened for business, and many drams were sipped in honour of 'the Polly'. The red and white flag of the Harrison line, under which the *Politician* had sailed, flies from a flagpole in the garden, while memorabilia salvaged from the ship, including portholes, a machete, Jamaican currency and, of course, a bottle of the famous whisky adorn the interior of the pub.

Compton Mackenzie wrote several other stories inspired by events that took place on our islands, the best known of which, after *Whisky Galore*, is probably *The Rocket Post*, the true story of a German scientist called Zucher who arrives on the Isle of Scarp in the Outer Hebrides with a plan to improve communi-

8

cations with the outside world by firing post bags in a rocket from the mainland. If nothing else, Mackenzie's stories prove that on islands reality can sometimes be more fantastic than fiction. (During a recent trip to Alderney in the Channel Islands this was brought home to me again when I came across the writer Elizabeth Beresford, creator of the Wombles of Wimbledon, who was working as a conductor on the island's train – a converted London underground train from the Northern Line. It was, I thought, a suitably surreal setting for the author of such whimsical characters as Uncle Bulgaria, Orinoco and, of course, Alderney.)

The islands around Britain have not only inspired many great stories but also served as havens and bolt-holes for writers, artists and musicians where they could shut themselves away to complete their works, free from the distractions of daily life on the mainland. If my own currently hectic work schedule allowed, I would love nothing more than to withdraw to one of my favourite islands for a few weeks or months and divide the days between putting pen to paper and enjoying the beauty of the land- and seascape.

George Orwell used to do exactly that. He wrote his masterpiece *Nineteen Eighty-Four* on the Scottish isle of Jura – a curious and incongruous setting, to say the least, for a writer to pen such a disturbing urban, anti-utopian tale. Biographers have struggled to explain Orwell's great attraction to Jura, but he certainly loved the island's stark, windswept beauty and prized the place because it was, in his own words, 'extremely un-get-at-able'. It was, you could say, his own little utopia. When I visited Barnhill, the remote farmhouse in which he lived, it was easy to understand why this place was once described as 'the loneliest house in Britain'.

My purpose on that journey to Jura was to see one of our great natural wonders, the legendary Corryvreckan whirlpool, which almost took Orwell's life when he was a young man. The Corryvreckan is one of the largest whirlpools in the world, and you can hear its awesome power from ten miles away. The collision of two tidal paths underwater pushes water upwards, causing enormous waves to rise up in the middle of the Sound of Jura with a violent vortex of water at their heart. Considered by the Royal Navy to be one of the most dangerous stretches of water in the British Isles, the whirlpool has claimed many lives through the years.

Orwell was sailing with his nieces and nephews when their boat was sucked into the Corryvreckan, ripping off its outboard. Fearing for their lives, he grabbed the oars and made a desperate lunge for land, but the boat was sucked under and the party was thrown into the frothing water. Fortunately, a local lobster fisherman had seen the drama unfold and quickly came to their rescue, unaware, of course, that he was saving the life of a man who would prove to be one of the nation's most cherished writers. I wasn't aware of this story until I visited Jura and heard it from one of the locals, which made me realize that there is such a vast amount to learn and uncover about our islands, even for an islophile like myself.

It is not just the island stories of the well-known that intrigue me but also those of 'ordinary' people. One in particular stands out in the memory and fuelled my hopes of one day acquiring my own offshore sanctuary. In 1965 Evelyn and Babs Atkins, two sisters from Yorkshire, were on holiday in Cornwall and discovered that a nearby island called Looe was on the market for £22,000. It had always been their dream to live on an island, so they sold everything to buy the place and ended up living there for the following forty years, selling pottery to make ends meet and help them live out their island fantasy.

Looe is a mile from the mainland and I went there several times to visit Babs, who lived in a tiny cottage surrounded by a thick wood until her death in 2004. She was a remarkably youthful, vital and resilient octogenarian who continued to live on her beloved island even after Evelyn passed away in 1997, with no one for company except a pack of wild dogs.

If you ever visit Looe you will understand Babs' unwillingness to leave it. The island enjoys stunning coastal views stretching from Prawle Point in Devon in the east to the Lizard Peninsula in the west, as well as an extraordinarily mild climate in which daffodils bloom before Christmas. The twenty-two-acre island, a former dropping-off point for smugglers, is a sanctuary for sea and woodland birds, and you can still see the remains of a Benedictine chapel built in 1139. Like most islands, Looe has its legends, but the story that Joseph of Arimathea once landed there with the child Christ might seem to stretch credulity to its limit.

Although the temperature very rarely drops below freezing, the winter storms rolling off the Atlantic can be so savage that sometimes the spray from the sea flies right over the island, which can be inaccessible by boat for days and even weeks. In earlier times communication with the mainland was conducted with flags and hand signals; I remember Babs' delight in showing me the whaleskin megaphone she and her sister had used. On another occasion, with thunder in her voice, she told me the tale of how some businessmen had visited them on the island with an offer to buy it from them for several million pounds. They planned to turn it into a theme park and, as an additional sop, they promised to erect two waxwork models of the sisters to greet visitors to a new offshore leisure centre. 'It must have been a little tempting,' I ventured. 'Did you ever consider accepting?' 'Did we hell!' stormed Babs, who, as promised, left the island to the Cornwall Wildlife Trust after her death.

Clearly I am not the only one out there enthralled by the mystique of islands. My friend the explorer and adventurer Bear Grylls and his wife Shara bought a nature reserve off the north coast of Wales, where they live in a lighthouse. Carla Lane, writer of the comedy series *The Liver Birds* and *Bread*, owns St Tudwal's East island off the Llyn peninsula, which she has transformed into an animal sanctuary. And journalist

Adam Nicolson memorably celebrated his beloved Shiant Islands in his book *Sea Room*. Such is the rage for owning your own island that there are now websites devoted to this burgeoning area of the property market.

The more I visited private islands, the more I wanted to own one for myself. About five years ago I found myself contacting estate agents up and down the country, asking to be put on their mailing lists so as to hear of any suitable properties that might crop up. There was, however, one obvious problem in acquiring my own island: they didn't come free and I barely had two spare pennies to rub together.

The first island I considered buying was Eilean Dubh or Black Island, one of the beautiful Summer Isles – which were given their name by local farmers who used the dozen or so islands to graze their livestock in the warmer months – off the Scottish west coast near Ullapool. The island was owned by a couple from Philadelphia, who read about my search for an island and offered me the chance to buy it; but in the end I had to face up to the reality that I simply couldn't justify the financial risk of burdening myself with a huge mortgage and a remote island, 'remote' and 'mortgage' being incompatible words. In 2001, however, I was in a healthier financial position, with a steadier income and a small flat of my own in west London, when the estate agent's particulars for an island called Ailsa Craig flopped through the letterbox.

Ailsa Craig is a highly distinctive island ten miles off

the southern Ayrshire coast in the Irish Sea. You can
see it from miles around as it soars a majestic 1,000
feet into the sky like an upside down pudding bowl. It
had once been the heart of an ancient volcano, but
today it is best known for the part it played in Britain
winning its first Winter Olympics gold medal for eighteen years in the 2002 Salt Lake City games, when five
Scottish women emerged as the surprise winners in
one of the curling events. Ailsa Craig – the name means
rock fairy in Gaelic – is the spiritual and geological
home of the curling stone; indeed, almost every curling
stone in use today is hewn from the unique, high-
quality 'micro-granite' found on the island.

This stunning island was on sale for the same price
as a modest one-bedroom flat in London, and with a
bit of creative juggling of the mortgage, I figured that
I could make it my own. My hopes were high when
the auction began, but sadly I missed the opportunity
to become King of Ailsa Craig and purveyor of curling
stones to the world's finest players, as I was outbid by
the RSPB, which has since turned the island into a
protected bird colony.

Acquiring my own island has proved to be every bit
as complicated and frustrating as I originally suspected,
even when the financial figures have added up. Another
island that has teased and tormented me over the years
is Eilean Mullagrach, also one of the Summer Isles.
Twice now I seemed to have the deeds in my grasp,
only to have them snatched away at the death.

On the first occasion I was pipped at the post by a

British family who bought the island, sight unseen, for £80,000. The family planned to use the deserted, treeless island as a very expensive family holiday camp-ground, but disaster struck on their first visit, when they were marooned on the island by a ferocious storm for more than a week without shelter or food. They were eventually rescued by the coastguard helicopter, but their experience had been so miserable that not long afterwards the island was put back on the market once again, much to my delight.

The 85-acre island is probably not everyone's idea of paradise – its Gaelic name means island of lumps and knolls, which, in truth, is a pretty accurate descrip-tion. Mullagrach has been designated a Less Favoured Area and is severely disadvantaged because there is no harbour, buildings or even a jetty. I didn't even own a boat that could take me to her but that, I convinced myself, was just a small detail that I would sort out at a later date. In fact all the island can boast by way of physical attributes besides its lumps and knolls are an underground spring and of course its coveted island status.

There is, though, something special about the Summer Isles, especially when you see them shrouded in the light of a setting sun, that lifts the spirits and warms the heart. There was no way I was going to miss out on Mullagrach a second time, even though the competition promised to be stiffer than ever. The estate agent Knight Frank had already received more than a hundred enquiries when I called but, promisingly, only one person had chosen to view the island, and that had been by seaplane.

Before taking the plunge, I was at least sensible enough this time around to accept that if I was going to buy an island I would have to be able to live on it. It would be pointless buying an island good only for sheep grazing and I didn't want to end up a laughing stock as the owner of Britain's most expensive field. A tent or yurt would be next to useless on this wind-battered outcrop, as the last owners had discovered to their cost. What I needed was planning permission; but when I called the relevant office I was told that the person in charge would not be back until deadline day, when all offers had to be tabled by noon.

In the meantime I forged ahead with my plans, seeking the advice of a local architect and increasing my mortgage so as to be able to meet the asking price. On the morning of the deadline I was unable to track down the planning officer until, finally, at 11.35 a.m., she picked up the phone. I quickly blurted out my plans for a building.

'Och,' she replied, 'I'm afraid we don't often grant permission for the Summer Isles as they fall within the Wester Ross National Scenic Area.' My heart sank. 'But,' she continued, 'depending on the application, we might consider a small residence so long as it was strictly in keeping with the environment.'

That was more than good enough for me. I had one foot ashore as far as I was concerned. I called the estate agent. 'I'll take it!' I screeched down the phone.

It was exactly 11.59 a.m., just seconds away from the close of bids.

'Wonderful, Mr Fogle, but I'm afraid we have received another similar offer,' the estate agent replied. 'It will now have to go to a sealed bid.' (Under Scottish law, the vendor is permitted to play off the prospective buyers against each other in an nerve-racking shootout.)

If I underbid, I risked losing Mullagrach, perhaps for just a few quid; if I was too keen, I risked paying way over the odds and being landed with an island worth less than I paid. It was an excruciating dilemma. I tried to imagine my rival and what was going through his or her mind as I wrestled with my own thoughts. Surely I was the only person in Britain foolish enough to buy a windblown, treeless, deserted island? I knew that the other buyer hadn't consulted the planning officer and felt certain that he or she hadn't visited the island. They can't have wanted it as much as I did – could they?

On the advice of my Scottish solicitor I increased my offer by 5 per cent, which seemed perfectly sensible. There was a noon deadline once again and I felt a rising sense of euphoria as the hour approached. While the lawyers and estate agent went about their business I couldn't bear the tension any longer. I walked around Hyde Park with my Labrador Inca, immersed in day-dreams of an island idyll. On my return, there was a message on my answering machine: 'I am sorry to have to inform you that you have been outbid.' And with those few words, my island dream was shattered once again.

Through such chastening experiences of trying to buy myself an island over the past few years I have learned a number of important lessons. Finding the right island, it seems, is a little like finding the right girlfriend or wife. For a start, you have to accept that many are out of your league and you simply cannot afford them or keep them in the manner to which they have become accustomed. Many others are no longer on the market, having been successfully wooed and courted by other suitors: these you can only stand back and admire from a distance.

You start your search for the perfect match with a long list of criteria: she must be attractive and charming, she should be a little distant but still hospitable and welcoming, she must belong to you and no one else, she must provide you with shelter from the occasional storm, she must present some kind of a challenge to sustain your interest and curiosity, and if she brings with her some useful material attributes and an interesting history that's all well and good too. You soon discover, however, that it's not a question of ticking all the right boxes, as you might at a dating agency, and that it's more a case of being in the right place at the right time.

When I decided to set out on a journey around offshore Britain, I had largely come to terms with the fact that, although there were thousands of islands out there, most of them were either already accounted for or wholly unsuitable for my purposes. But maybe, just maybe, I would come across my own little piece of utopia for sale, or a magical isle on which I could at least settle.

My adventure was probably not so much an epic property search as a grand and extravagant excuse to indulge in my lifelong passion for islands and my curiosity about those who live on them; an opportunity to meet, explore and understand my own people, my neighbours. After many years spent travelling to every corner of the globe and visiting remote dots on every page of the atlas, it had dawned on me that there was a whole world of islands and communities sitting on my doorstep, each with their own distinct character, history and beauty.

The British have a very curious relationship with their islands. Mostly they exist only dimly on the fringes of people's consciousness and yet when one of them comes up for sale, or finds itself convulsed by some kind of controversy or scandal, the newspapers devote pages to the story. There is a part of us that is enthralled by island life, but another that is fearful and suspicious. Part of my mission, this journey through the heart of our offshore community, was to get to the bottom of this ambivalence. What exactly is it about our islands and their communities that so fascinates and has such a strong hold over the British imagination?

# Sark

'Welcome to Grand Cayman!' said Andrew in his distinctive pidgin drawl as we made our way through the smoky, spicy aroma of West Indian cooking to the welcoming beat of Bob Marley and the Wailers. A seemingly unremarkable greeting had I just touched down in Kingston, Jamaica, or Port of Spain, Trinidad, but as it happened I had just arrived on Guernsey in the Channel Islands, roughly four thousand miles from the Caribbean.

I was passing through Guernsey en route to Sark, one of the most beautiful and extraordinary islands in the northern hemisphere, and I had stumbled on one of the world's most bizarre gatherings: the NatWest Island Games. Triumphantly billed by the organizers as the third biggest sporting event in the world – bigger even than the winter Olympics – the Games would see over 2,200 athletes plus coaches and officials descend on the Channel Islands for the biennial athletic battle between some of the most remote and little-known island communities on the planet. Guernsey was crawling with tracksuits that bright summer morning as the sporting cream of the oceans poured in from Aland, Bermuda, Alderney, the Cayman Islands, the Falkland Islands, the Faroe Islands, Frøya, Gibraltar, Gotland,

Greenland, Guernsey, Hitra, Isle of Man, Isle of Wight, Jersey, the Orkney Islands, Prince Edward Island, Rhodes, Saaremaa, St Helena, Sark, Shetland and Ynys Môn (Anglesey to you and me). While many were familiar to me from my various island odysseys over the years, to my shame as a proud islophile, I had barely heard of half the others. Saaremaa and Frøya sounded more like Scandinavian breakfast cereals than islands with netball teams.

The Island Games began in 1985 on the Isle of Man as a crafty wheeze to boost sport and tourism. Only six islands from around Britain and Scandinavia were invited to compete in six events over a week. It was such a success that the event has expanded rapidly with each Games since. The definition of the word island, however, had clearly been stretched a little to allow others in on the fun. Can Gibraltar and Greenland truly be called islands? Unless there has been a dramatic acceleration in global warming and southern Spain has been severely flooded since my last visit, I am not aware that Gibraltar is geographically independent of the mainland. And as for Greenland, can a place bigger than continental South America really be classified as an island? If so, why not invite Great Britain and Australia while you're about it? The reason, according to Geoffrey Cortlett, the Pater Ludorum or Father of the Games, is that the get-together is as much to do with political isolation as geographical, and in an open-minded and charitable bending of the rules, both places have been awarded small island status and welcomed into the family fold.

The event is largely ignored by the rest of the world, but for those taking part it has become one of the highlights of the calendar, and live coverage of clay-pigeon shooting, judo, basketball and many other 'Olympic' sports is beamed live to various outcrops around the globe, from the Faroes to the Falklands. For me, this was an unexpected boon, as I had only recently returned from some of the more remote islands taking part. When I arrived back in Britain I had wondered whether I would ever again come across any of the characters I had met on my travels, and so it was a great thrill when I walked through the competitors' tented village on Guernsey and heard a voice behind me say, 'Hey, Ben, how are you doing?' The voice belonged to a Falkland Islands tracksuit. It was Andrew from Stanley. 'Hey, what are you doing here?' asked another competitor in a St Helena sweatshirt. It was my taxi driver from Georgetown on the South Atlantic Isle. Small islands, small world.

From this one week in July, the gaze of the world's smaller islands would fall on this little bailiwick in the English Channel, their athletes competing for gold and glory, to be hailed as heroes on their return, win or lose. Traditionally each island brings a bottle of island water and one of the athletes pours it ceremonially on a fountain – the Games' equivalent of the Olympics' torch ritual. Island water is not just a tonic but a door to the world.

The opening ceremony was fittingly focused on

the harbour in the capital, St Peter Port, and involved each competing nation or dependency sailing in
aboard boats festooned with its colours, flags snapping
proudly in the breeze. St Helena was given a rapturous
reception after the revelation by a man on the tannoy
that the team had spent three weeks travelling from
the remote South Atlantic island. Later there was an
audible gasp at the announcement that Hitra, a northerly Norwegian island, is now connected to the mainland by one of the world's deepest tunnels, 2,842 feet
below sea level. Many believe that isolation gives
islands their distinctive character, which can only be
diluted by connection to the mainland by tunnels and
bridges.

It was an unexpected thrill, as I set out on my own
island adventure, to find myself mingling with such an
eclectic mix of islanders. They all shared a common
bond, a kind of mutual understanding to which non-
islanders such as myself could never be entirely privy.
There was no pomp that evening, just ceremony and
fun as the men and women from Prince Edward
exchanged tales and drinks with their island cousins
from Shetland. It was rare to experience an atmosphere
of such innocence and delight as these very different
people from every corner of the world were brought
together to celebrate the one thing they all had in
common: their island identity. I felt as if I was a cub
scout at a meeting for grown-ups.

I would like to have stayed longer but my journey
demanded that I leave the festive excitement of

Guernsey for the peace and tranquillity of neighbouring Sark. Only seven miles separate the two islands but
the contrast between the two places could not have
been greater.

When I had flown into Guernsey on the short flight
from Southampton, I had been struck by how built-up
the island looked from the air. On my various travels
I had grown used to the small, scattered communities
of islands with single-storey, low-rise buildings, and
basic roads and amenities, but Guernsey has a positively cosmopolitan, moneyed feel to it. It was not
always that way, of course, but over the years its special
tax status has led to a boom in banking and corporate
activity, as well as many of the people who live there
having a very glamorous lifestyle. There are a lot of
flashy houses on the island accommodating the many
millionaires who live there but you also come across
sprawls of unexceptional houses, villages and communities like those you would expect to find anywhere
in southern England. Convertible sports cars speed
along roads lined with familiar high street names, bars
and restaurants, while in St Peter Port international
banks and 'conference' hotels seem to spring up from
every corner.

In my mind's eye I had expected to find a largely
rural, old-fashioned world but instead I found a mini-
version of Bournemouth or Brighton, more cocktails
and Ferraris than cows and flowers. I felt a little uncomfortable when I went out for the evening in
St Peter Port in my scruffy jeans and Guernsey while

all the other young people were kitted out in chic designer clothes. (I must have been the only person under the age of forty on Guernsey wearing the jumper to which the island has given its name. I had even made a last-minute flying visit to the factory to stock up on my favourite item of clothing.) Guernsey is still a very beautiful place, and you can drive out along winding narrow lanes through a patchwork of fields and find yourself a quiet corner of a beach or a cove. But with a population of 60,000 people, a land mass of 25 square miles, a thriving business culture and a glitzy style of life for many of its people and visitors, it felt more like the mainland than an island to me.

Sark is a world apart, not just from Guernsey but from anywhere else I had visited. It may be a tax haven, but it is also a haven of tranquillity and stunning beauty. I was instantly smitten by it. The smallest 'independent' state in the British Commonwealth and one of the last bastions of feudalism in the world, this little pinprick on our maps is just one hour by boat from the bright lights and bustle of Guernsey, but like another planet. It is just three miles long and a mile and a half wide, but its size belies a rich and extraordinary history and an abundance of natural charm.

Only 610 people live on Sark, and it is perhaps not surprising for an island that stands 25 miles from the Cherbourg peninsula of Normandy and over eighty miles from England that its atmosphere and culture have a distinctly French flavour to them. Most of the place names are French and the islanders still speak the patois Sercqulaise. The island is governed by the Seigneur of Sark. There have been twenty-two seigneurs since 1565, when Elizabeth I enacted the island's laws, many of which remain unaltered. In that year Helier de Carteret solicited Elizabeth for a charter to colonize Sark with forty families from his native Jersey in what proved to be a mutually beneficial arrangement: de Carteret became lord and master of his fiefdom, while for England and the Queen it provided an opportunity to end centuries of anarchy, piracy and smuggling in the region. De Carteret was granted his wish on the condition that he had a force of forty armed men to keep the island clean and secure from undesirable elements. For the following 300 years the island remained the property of the de Carteret family. The original law requiring forty men to be kept under arms remains on the statute books, although it is rarely invoked. When Helier de Carteret was officially appointed Seigneur he divided the island into forty separate tenancies, one for each of the families he had brought with him. These tenancies were declared indivisible and have stayed so for 440 years.

It was a beautiful summer's day as our little ferry headed out from St Peter Port for Sark with its cargo of two or three dozen people. It wasn't difficult to tell who was visiting the island and who lived on it. The day-trippers wore backpacks and bright cagoules and stood outside in the boat to admire the view, while the locals sat inside, their heads buried in newspapers and books, barely looking up until it was time to disembark.

As we approached the island I was struck by the enormous cliffs that encircle its coastline like the fortifications of a medieval castle. I could only guess as to the landscape at the top. As I looked at the island rising sheer out of the water for about three hundred feet with no sign of roads or beaches, it was easy to understand how it came to pass that Sark remained cut off from the outside world. There is a story that the Lords of Admiralty once sailed around the island with a plan to build some defence works but quickly reached the conclusion that nature had already done their job for them.

It took centuries to construct a safe harbour and the three that you find there today – Creux, La Maseline and Havre Gosselin – are all very small and capable of accommodating only a few boats at a time. Tunnels had to be blasted through the rocks in order to access the island from the landing places. Our ferry docked in the Creux harbour, which is reputed to be the smallest in the world, and we were met by the most modern and sophisticated form of transport on the island – a trailer pulled by a tractor, known affectionately by the locals as the 'toast rack'. As I settled down in my seat with the other visitors, I was amazed to see that the locals were heading on foot up the single track. Sark's inhabitants have big lungs and strong calf muscles.

The 'toast rack' is the only motorized transport vehicle on the island and as it puffed its way through the little tunnel, up the steep hill and into the main street – or the Avenue as it is known – it quickly became clear that I had landed on a very unusual island. The 'town' is not so much a settlement as a delightful scattering of quaint stone houses and little shops. There are no cars, no paved roads and no street lights. Horses attached to miniature carts were tethered to wooden railings as if in a scene from the mid-West, while a group of people sat outside a café in the glorious morning sunshine, sipping coffees and teas, and locals came and went from the tiny old-fashioned post office with the day's papers. It was a charming scene, but there was nothing of the chocolate box about it: everything struck me as truly authentic rather than geared to impress and entice visitors.

Sark is like an island that history forgot. I felt as if I had been transported back in time to a Victorian or Edwardian fantasyland. I was acutely aware of the silence. There was no purr and rumble of car engines, only the clip-clop of horse hooves as they trundled their little traps up and down the gravel roads. Gone too was that most modern plague, the mobile phone ringtone. Sark remains defiantly 'signal free'; the only sounds you hear are natural.

I was staying at the relatively upmarket La Sablonerie in an area of the island known as Little Sark. To get there I hired a bicycle and headed off down a gravel track into one of the most beautiful panoramas you could hope to lay eyes on, a colourful fusion of land and sea bathed in warm summer sun. The countryside was completely unspoilt by modern life and its diversity remarkable considering its size: dramatic valleys, lush woodlands, green fields, swathes of wild flowers and the odd sandy beach all vied for its limited space. Primroses, dog violets and gorse lined the pathways and the fields in a blaze of colour, giving out wonderful scents.

One of the curiosities of the island is that all the signposts measure journeys not in terms of distance but by the length of time they take. Everywhere you go, the little figures of a walking man, a bicycle and a horse and cart guide you to your destination.

Finding La Sablonerie was easy, partly because you are never too far from any one place on Sark and partly because it is one of the island's most famous landmarks. A long, low white building festooned with flowers and hanging baskets, the 400-year-old hotel is the epitome of understated charm, set in the grounds of a farm and gardens, which between them provide almost all the ingredients for the restaurant's menu. The hotel is run by Elizabeth Perrie, and even before I had set foot on the island I had heard much about the warmth and generosity of her hospitality. (The family own four of Sark's forty tenements – virtually the whole of the area known as Little Sark.)

I sat in the English garden surrounded by a riot of colourful flowers, as Elizabeth, French of face and English of intonation, plied me with glasses of Pimms and tales from the island's unique history. Within an hour of my arrival my head was swimming, and when I remounted my bicycle to tour the island in the afternoon I could have done with a set of stabilizers to stop me wobbling about.

Little Sark and Greater Sark are connected by a stunning narrow, highly vertiginous isthmus called La Coupée, which is just nine feet wide at its narrowest point and drops 300 feet to Convanche Bay on one side and to a beautiful sandy beach called La Grand Grève on the other. This great expanse of golden sand, caves and rock pools is one of the few accessible stretches of the island's cliff-lined coast.

Before protective railings were erected on La Coupée in 1900, small children had to crawl across it on their tummies to avoid being blown over the edge, and even today it is mildly alarming when there is a strong breeze blowing off the Channel – even more so when you are on the business end of two or three strong glasses of Pimms. Covering the entire length of the isthmus there is a narrow concrete road, which was constructed in 1945 by German prisoners of war under the direction of the Royal Engineers, a small memento of retribution for the islanders following their five years of occupation in the Second World War. Those forty muskets decreed by Elizabeth I were of little use when pitched against the might of Hitler's military machine.

The occupation of the Channel Islands was probably the most painful episode in their long history. German forces first landed on Guernsey in the summer of 1940 and within a day or two all the islands had fallen under their control. Islanders were dispatched to prisoner camps, others were shot for offering resistance, while some openly collaborated with their invaders. Sark was heavily mined and wired after a succession of British commando raids. The Germans quickly set about building elaborate fortifications, convinced that they were just a few weeks or months from launching an invasion of southern England. Gun emplacements and observation towers were erected all across the islands, but that was as far as their conquest of the United Kingdom went, for the pilots of the Battle of Britain put an abrupt halt to Hitler's ambitions. By 1944, following the Allied landings across the water in Normandy, the occupying forces were cut off from France. Food and basic supplies for soldiers and civilians alike were quickly exhausted and many were forced to scavenge for food, with cats, dogs and dead fish all finding their way on to the plates of a near-starving people.

When Allied forces landed to reclaim the islands, the occupying forces were said to have been as relieved as the locals. While Guernsey and Jersey celebrate their liberation from German occupation on 9 May, Sark had to wait a further twenty-four hours before the islanders could start re-building their lives. Every five years the people of Sark celebrate their liberation day with a parade through the main 'town', but the war is a passage of their history that most residents would prefer to forget.

Today there is certainly no shortage of food or money on Sark, or any of the other Channel Islands for that matter. Sark is home to some seriously rich people, but unlike Jersey and Guernsey, there is very little obvious display of wealth. Everyone, from the millionaires to the farm workers, looks remarkably similar. They all wear sensible island clothes and Guernsey jumpers, and they all get around by bicycle. You can often tell someone's status by the car that they drive but that's not so easy with bicycles.

As I continued my cycle ride around the island, I was struck by the lack of ostentation of its people and the homes in which they live. But there may be a very sensible reason behind the down-to-earth, no-frills appearance of the islanders. On Sark there is no income tax, no VAT and no death duties, but instead there is a revenue rating system called Visible Wealth Tax, which is set by the Douzaine, a committee of twelve people elected from the island's governing council, which is called the Chief Pleas. The Douzaine meet annually to assess the wealth of each resident and to impose tax accordingly. In practice, the tax bill of most islanders rarely exceeds £200. The island also raises money by a form of poll tax, charged to all visitors to the island, as well as through levies placed on alcohol and tobacco. Although there is constant talk of changing the ancient laws, self-governed Sark, for the time being at least, remains a fiscal paradise the rest of us can only dream about.

Employment on Sark, apart from some dairy and sheep farming, is largely seasonal, and the island is heavily dependent on the tourist industry. They say there is no unemployment on the island, but there is also no benefit system. The Douzaine dispenses Poor Relief to those who are deemed to be deserving cases.

For years, one of the most profitable lines of work for some of the islanders (if 'work' is the right word for it) was something that became affectionately known as the Sark Lark. Until the Bailiwick of Guernsey was forced by mounting pressure to tighten up lax financial regimes of Guernsey and Sark about five years ago, it was rumoured that virtually every adult on Sark – and many on Guernsey – was the nominal director of several offshore companies. The directors often knew virtually nothing about the companies they represented, but most felt little compulsion to ask so long as they continued to receive handsome cheques for doing precisely nothing, except attend the odd business lunch. Some islanders, it is said, were the directors of over two thousand firms and estimated to have earned over £100,000 just for living on Sark. Not bad work if you could get it, but it's probably not best to raise the issue over a drink in the local pub.

From La Coupée I continued my cycle journey to the island's other landmarks. I headed to the intriguingly named Le Beq du Nez on Sark's northernmost outlying rock, known to the older generation of Sarkese as the Oystercatcher's Rock. It was here in 1839 that the island's Seigneur was drowned after putting out to sea in rough weather. His close friend the Vicar of Sark witnessed the disaster and developed such a fear of the sea as a result that he refused to leave the island for the remaining thirty-eight years of his life.

I visited that unfortunate vicar's tiny church, situated on the fittingly named Rue du Sermon, where, in keeping with Sark's finest feudal traditions, the pews are 'owned' by families on the island, who pay the church two pence annually for each seat. I cycled on past Le Pot, so named because of its distinctive shape. Le Pot is the former site of a very short-lived 'silver rush' in the nineteenth century – a somewhat hairbrained venture that ended in disaster when the area was flooded and the bankrupt Seigneur was forced to sell his title to cover his losses. Today, the odd chimney and some scattered ruins, including a crumbling jetty, are the only evidence of Le Pot's brief, ill-fated existence.

All this cycling was making me hot and so I returned to La Coupée and headed down the steep steps to the beach at La Grand Grève to cool off by its stunning turquoise waters. The waters around Sark are the clearest in the Channel Islands, the locals say, and they contain a wealth of marine life. Dolphins and porpoises can often be seen off the coast and there are also rich pickings for professional and amateur fishermen alike, including the dazzling cuckoo wrasse, grey and red mullet, conger eel, bass, pollock, bream, mackerel, flatfish, dogfish and whiting as well as lobster and crabs around the island's rocky coastline.

The island used to be popular with both pirates and

smugglers and today you can visit the caves and inlets along the cliffs where they hid their ill-gotten gains. The Boutique Caves were once a popular hiding spot for booty and contraband because they are accessible only at low tide and made up of a complex series of tunnels and chambers.

My final port of call was La Seigneurie, one of the few houses on Sark with any pretensions to grandeur, with a central watchtower and impressive walled garden complete with vast greenhouse, maze and dovecote. The present house was built on the site of the sixth-century monastery of St Magloire, but it has been substantially altered and extended over the years. Today it is home to former British Aerospace engineer Michael Beaumont, the island's twenty-second Seigneur. He inherited Sark from his grandmother, Dame Sybil Hathway, who is remembered with great affection for helping the islanders through the trauma of their occupation during the war.

The Seigneur, by all accounts, is an extremely keen horticulturist, who spends most of his time tending his beautiful gardens. The place had an ethereal, otherworldly feel, I thought, as I parked my bicycle by the gate and wandered a little nervously down the drive (or in carless Sark's case, the 'walk') to meet the constitutional head of this extraordinary little sovereign state.

Michael holds the island in perpetuity from the monarch in return for taking an oath of allegiance, and paying an annual rent of 'one twentieth part of a knight's fee' – a cheque for £1.79 to be exact, which he writes out to the Queen's representative on Guernsey. The Seigneur is entitled to a *treizième*, or one-thirteenth, of the sale price of all property traded on the island.

Sark's administration and way of life are intimately bound up with archaic laws and conventions, many of which remain unchanged. The Chief Pleas (its governing council) meets three times a year and is composed of the island's forty landowners, the Seneschal, the Greffier, the Treasurer, the Constable and the Prévôt (whose main job is to feed any prisoners in the tiny, one-room jail), and presided over by the Seigneur.

The Seigneur has *droit de colombier*, the sole right to keep pigeons, and is the only person allowed to keep an unspayed bitch. So if you happen to be a pigeon fancier or dog breeder, you are probably better off avoiding Sark. Until 1852 no one on Sark could marry without the Seigneur's permission.

I had been expecting to find myself greeted by some gruff old tyrant but was pleasantly surprised to find a gentle, mild-mannered man more intent on wielding a pair of garden shears than his bizarre constitutional rights. This tall and slender seventy-year-old could scarcely have been less imperious in voice, manner and gesture as he proudly walked me around his highly acclaimed gardens.

'I still collect tithes,' Michael said, smiling shyly. 'Most people pay in cash now, although I do occasionally collect wheat. I remember my mother collecting tithes in chickens. The tenants would always give her the scrawniest and toughest old birds. I'm just an elaborate Lord of the Manor and whereas we used to be two a penny, now I'm the only one left,' he added, cutting off a dead head from a rose bush with his shears. I quickly understood that Michael had a genuine fondness for Sark's traditions and heritage, but he feared for the island's future status. 'It won't last for ever,' he said with a sigh. 'The European court will get their teeth into us and that will be it. We can't live in the sixteenth century for ever.'

It is true that some of Sark's laws seem truly odd to outsiders. No aircraft, for instance, may fly over Sark below 2,000 feet except in an emergency or for a VIP visit (the Queen, apparently is a regular visitor). Also under Sark's fishing laws, 'it is forbidden to take fish of any kind whilst being totally or partially submerged, and breathing with the aid of a breathing apparatus or wearing a face visor.' Another ancient law insists that female members of the governing council must keep their heads modestly covered. Divorce is banned. Until 1975 women on Sark had no rights whatsoever. It is said that a man could beat his wife, but if he drew blood he would find himself in trouble. Technically women were not allowed a bank account, though many ignored this; and when a woman married, all her property automatically passed to her husband. All this, however, changed at the instigation of Michael's mother, the formidable Dame, who shortly before her death in 1974 declared that women on Sark should have the same rights as women everywhere in the West.

Though democracy, meritocracy and human rights, which are taken for granted elsewhere, have been slow to reach Sark's shores, that was all part of the island's quirky charm, I thought, as I cycled back along the narrow tracks and past picturesque weatherbeaten cottages. I continued on my way to Pilcher's Monument, a stone finger erected to mark a Victorian tragedy in which five men perished one foul night in October 1868. One corpse was found at the Havre Gosselin harbour, from where their ship had set sail for Guernsey, another at the extreme north of the island, a third in Gouliot caves close to the harbour and a fourth, that of J. Pilcher, a London merchant, a few months later on the Isle of Wight. The sea had stripped the flesh from their skulls and hands, but their clothing remained intact. The fifth corpse was never found. Pilcher's widow commissioned the granite memorial to stand above Havre Gosselin on top of the towering cliffs, pointing a warning finger skywards for passing ships.

It was an eerie place, and I soon headed on a few hundred yards along the western coast up to Gouliot headland, where there is a splendid view out across the sea to another unique island, Brecqhou. This curious rock of an island rises from the sea like a granite fortress surrounded by a moat of raging water. It squats on the other side of the Gouliot Passage, a dangerous funnel where the rip tide swirls through at up to 15 knots. The sea boils against the foot of the cliffs of Pointe Beleme, a series of fissured rocks rising 240 feet into the sky.

'Beautiful, isn't it?' announced a gorse bush.

'What brings you here?' asked another.

It seemed a rather unlikely place for a biblical experience, but I had heard tales about the fairies that earlier residents believed inhabited the island. I followed the voice to an easel with a bearded man behind it.

'Hi, I'm Graham,' he said, smiling and offering me his paint-splattered hand.

'And I'm Rosanne,' announced the other bush.

'Is that *the* island you're painting?' I asked conspiratorially.

'Yes. They say it's patrolled by dogs and guards,' whispered Rosanne as she dragged her brush across the canvas. 'No one goes there. It's strangely beautiful, though, isn't it?'

I looked at the great stone behemoth rising from the centre of the island, a vast medieval castle of questionable aesthetics, and wondered whether beautiful was the most fitting way to describe it.

Brecqhou is the secretive home of millionaire twins Sir David and Sir Frederick Barclay, who bought the 80-acre island in 1993 for £2.3 million. Little is known about the Barclays, who started life with very little money but made a fortune in property and shipping and whose vast and powerful portfolio of possessions now boasts the Ritz, the *Scotsman*, the *National Enquirer* and the *Daily Telegraph*. The twins are estimated to be worth between £500 million and £1 billion, and said to give millions away to charity every year. Only one photograph exists of the pair, who are renowned for their reclusiveness and insistence on privacy. It is claimed that the only way of telling them apart is by which side they part their hair, each parting his hair on a different side.

They hired Prince Charles's favourite architect, Quinlan Terry, to build their baronial fortress at an estimated cost of between £80 and £120 million. It required 2,400 boat trips to transport the 70,000 tons of building material, including 12,000 tons of Spanish granite. 'It was a worksite for years,' explained Graham, who went on to tell me that a thousand men, many of whom had come straight from the Channel Tunnel project, had worked around the clock to build the castle, complete with two swimming pools. It is estimated that the building created employment for roughly five thousand people in total. 'They have their own flag with their own coat of arms,' whispered Rosanne, as if there were spies listening in the gorse bushes. I could just make out the flag through my binoculars, fluttering from a mast above one of the turrets. I could also see a helicopter with its rotors tethered.

The brothers have created their own little unofficial sovereignty, complete with its own stamps, and they have caused quite a stir in Sark's corridors of power by challenging the ancient constitutional arrangements which decree that Brecqhou falls under the jurisdiction of the larger island. The brothers claim democracy on their side and have lobbied the Lord Chancellor to intervene in the smouldering dispute, which has potentially huge implications for Sark and its people. To someone like me who is not an expert on constitutional or international law, it is difficult to argue with the Barclays' claim that the proposed reforms do not go far enough and leave too much power in the hands of unelected hereditary landlords.

Sir David Barclay feels so strongly about the issue that he recently broke the habit of a lifetime and gave a rare interview to the *Observer*, in which he was quoted as saying, 'Sark's feudal lord and his colleagues have enormous power. This will continue under the new constitutional changes, because of potential intimidation in a small community where people are often afraid to speak their minds.'

Naturally, the Barclays' dispute has created quite a stir on Sark, where things had gone largely unchanged and unchallenged for the better part of half a millennium. A mock castle costing £120 million is one thing, but independence is quite another. Like Queen Victoria, the islanders are not amused.

One of the issues of the case was inheritance. Under the Sarkese law of primogeniture, all property must be left to the eldest son, but the Barclay twins want to leave the island in trust to their four children, one of whom happens to be a girl. They sought to prove that Brecqhou comes under the jurisdiction not of Sark but of Guernsey, where, as in mainland Britain, they could include all their children in their bequest, but the Sark authorities maintained that the effect of the brothers' civil action was to undermine the constitutional link between the two islands and that that, in effect, amounted to a unilateral declaration of independence.

I knew only a little about Brecqhou before I stumbled into those gorse bushes, but I was intrigued, to say the very least, by the tales of the secretive twins, their dogs, guards and courtroom battles.

'Has anyone visited the island uninvited since the Barclays moved in?' I asked my gorse bush artist friends.

They both shook their heads in unison and let out a long 'Nooooooo'.

An idea was starting to form in my head.

Graham told me the cautionary tale of one trespasser who bore the brunt of the twins' wrath. In 1995 a journalist called John Sweeney landed uninvited on Brecqhou to make a BBC2 television programme about the island and the castle. The Barclay twins were livid and took Sweeney to the High Court, suing him for £135,000 for an article that subsequently appeared in a Sunday newspaper. They claimed that the article infringed their privacy because it named the road on which they lived in Monte Carlo 'without legitimate reason' and that it could make members of their family targets for kidnappers. They further contended that an accompanying drawing of them infringed their 'absolute right to their own image'. They lodged a complaint with the Broadcasting Complaints Commission and

even took Sweeney to the French courts for an interview he conducted on BBC Radio Guernsey, claiming
further damages of £108,000 for 'significant emotional
distress'. Luckily for Mr Sweeney, they lost the case.

But Sweeney's troubles didn't end there. The
Barclays persevered and sued the then BBC Director-
General, John Birt, for criminal libel. They accused
Sweeney and Birt of 'public slander' under French
criminal and civil law at a court in St Malo in Brittany
because the broadcast was picked up on the French
mainland. Sweeney and Birt were found not guilty, but
the brothers appealed and had the ruling overturned.
Sweeney was relieved of a rather more modest 20,000
francs (roughly £2,000) but the court rejected the
Barclays' demand for damages against Birt.

The details of my invasion plan were starting to take
shape in my mind as Graham explained the ins and
outs of the legal wrangle to me. I have always enjoyed
a bit of an adventure on the high seas, but was I really
prepared to risk a repetition of such wrath and, more
to the point, could I afford such high jinks? After all,
this was a family that was prepared to challenge not
only the BBC but the constitutional authority of the
Queen herself.

The law of trespass on Brecqhou is a little bit odd,
to say the least. Among the many strange Sarkese laws
is the old Norman custom of *clameur de haro*, a legal
device which exists today only in the Channel Islands
by which a person can effect the immediate cessation
of any action he considers to be an infringement of

his rights. At the scene he (for example, a Barclay twin) must, in front of witnesses, fall to his knees and recite the Lord's prayer in French and then cry, '*Haro! Haro! Haro! A l'aide, mon Prince! On me fait torte!*' ('Hear me! Hear me! Hear me! To my aid, my Prince! Someone does me wrong!') On hearing this, the wrongdoer (for example, me) must cease his activity and the 'criant' (for example, the Barclay twin) has twenty-four hours to register his complaint with the Greffier's office. If found guilty, the perpetrator risks punishment, which traditionally has taken the form of a fine or a night in the deepest dungeon in Guernsey. If the criant is found to have cried '*Haro*' without good reason, it is he who faces a night in the cells.

The last recorded *clameur* took place in June 1970 in a tiff over the construction of a garden wall, and I was gambling that the brothers wouldn't be waiting for me with a French dictionary on the other side of theirs. Besides, I figured that a night in a Guernsey dungeon would be quite cool. I returned to the 'town' to find someone willing to give me passage across the short stretch of water to the 'forbidden isle', but perhaps unsurprisingly, the locals weren't exactly queuing up.

Having exhausted all other options, I resorted to plan Z and made my way to the island's little beach shop, purveyor of fine buckets, spades, beach balls and children's inflatable boats. I had decided to launch a full-scale invasion on Brecqhou in a Mr Incredible dinghy. It was not exactly the high-tech stuff of Bond movies, but it was the only 'boat' on sale in the island's gift shop and therefore offered me my only chance.

I cycled back across La Coupée, scrambled down the cliff to the rocky beach below, unrolled *Mr Incredible* and puffed air into it, turning blue as my lungs struggled to fill my invasion craft. I had to admit that it looked very small and unimpressive indeed as I strode into the shallows with it tucked under my arm. I waded out as far as I could and then carefully wedged myself on board with my knees up by my chin, my hips straining the walls to bursting point and my bottom scraping along the rocky seabed. I felt like a champagne cork about to pop. Writing on the dinghy informed me that the craft was suitable for children up to eleven years old.

Undeterred, I manoeuvred the blue and red plastic oars into their locks and rowed out into the clear blue Channel waters. My elbows jarred against my stomach as I tried to heave my way to Brecqhou and I can't have travelled more than about a hundred yards when my hopes of getting there literally began to deflate. Pssssssssssssss. *Mr Incredible* was clearly not as incredible as he might have liked to think. Water started spilling over the side as the dinghy's walls collapsed. I was sinking. My craft had been mortally holed below the waterline (or rather, my big bum had scraped against a sharp rock, which had punctured it) and before I could say '*Haro! Haro! Haro!*', it was '*Merde! Merde! Merde!*' and I found myself wallowing waist deep in water, clutching on to my flaccid dinghy, shipwrecked before I had even departed.

I dragged the craft back to shore and sat on the rocks, exhausted and feeling ridiculous. When the light began to fade, I put the useless *Mr Incredible* into the bin where he belonged, put my saddlesore, rock-scraped behind back on my bicycle and pedalled off to La Sablonerie. It had been a long day on this enchanting island. I sat in the garden with a glass of Pimms to hand as I watched the setting sun leave a giant smear of orange across the Channel. I may never make it to Brecqhou, but I had been charmed by Sark, a small slice of island heaven right on our doorstep.

# Caldey, Bardsey and the Skerries

## *Caldey*

The variety of the islands in our waters, both in their physical differences and in the purposes to which they have been put, is remarkable. Over the centuries they have been used as lighthouses, military bases, smuggling centres, tax havens, privately owned fiefdoms, fortresses and castles, fishing businesses, wildlife sanctuaries, holiday resorts, hippy colonies, refuges from persecution and hiding places. The list goes on, but if there is one specific purpose to which they have been put more than any others, it is that of religious and spiritual retreats. After all, if it's peace and solitude you need, what better place is there to cut yourself off from the world, both spiritually and geographically, than an island?

One thinks immediately of St Columba on Iona, and St Aidan and St Cuthbert on the holy island of Lindisfarne, who together did more than anyone else to bring Christianity to Britain during the Dark Ages. Lindisfarne, a tidal island off the Northumbrian coast, remains a pilgrimage centre for Christians to this day, receiving over 500,000 visitors a year.

Scotland, too, has its own holy island, off the coast

of Arran in the Firth of Clyde in the south-west of the country. At the end of the sixth century an ascetic Irish noble, Molaise, abandoned his home to become a hermit on this beautiful island but, thanks to his ability to work miracles, his fame spread and St Columba dispatched him to Rome. The island subsequently became a place of pilgrimage for Celtic Christianity and in the thirteenth century a monastery was built at its northern end. A decade ago it was bought by a group of Tibetan Buddhists, whose aim has been to preserve its remarkable spiritual heritage, environment and wildlife.

Further north, on the Orkney island of Papa Stronsay, is the Golgotha Monastery, which dates back to the seventh century but has only recently been restored to its original purpose after 700 years in abeyance. This most northerly Christian monastery ever founded is today home to a community of Transalpine Redemptorists, who are entirely self-sufficient.

At the other end of the country there is St Michael's Mount, a spectacular islet that soars 258 feet out of Mount's Bay in the far south-west corner of Cornwall. Its creation is attributed to the giant Cormoran, whose heart, legend has it, is buried there. There is evidence that several centuries before the birth of Christ the place became a port dealing in tin. The Mount acquired its name in the year 495, when the archangel St Michael is said to have appeared on the rock to a group of fishermen. A chapel built there became a shrine in the Middle Ages, making it Britain's oldest pilgrimage centre.

Perhaps the strangest 'hallowed' island around our coast lies a few miles off the Kent coast, where the River Medway melts into the Thames. A muddy splodge measuring nine miles by five, speckled with docks, houses, caravan parks and farmland, Sheppey differs from the average holy island in having a population of 40,000 people. The abbey at Minster was founded in 670 by a Saxon queen with possibly the most suggestive name ever invented, Sexburga, who chose the location because of the healing waters that ran underground. The abbey became a stopping-off point for pilgrims on their way to Canterbury, and the waters acquired a reputation for enhancing potency, which is perhaps why Henry VIII and Anne Boleyn honeymooned on Sheppey and Horatio Nelson and Lady Hamilton courted there. Today, though, you will do well to find much evidence of active religious devotion: where once you would have found cowls and vestments, you are now more likely to see tracksuits and Burberry baseball caps.

There are dozens of smaller, less well-known islands dotted around the coastline that have been colonized by religious communities over the centuries. Most of them have long since been swallowed up by history, and only a few ruins and the odd historical reference remain to remind us of their existence; but there are still a handful of communities that survive, and even thrive, today. I am not a greatly religious person – I'm an agnostic, I suppose – but I have always been interested in religion and the part it has played in shaping our heritage and history. As I continued my clockwise journey around Great Britain, I took the opportunity to visit two of these religious outcrops, which jut out of the Irish Sea off the coast of Wales: Caldey in the south and Bardsey in the north, both of which today are home to fascinating, if very different, communities.

Caldey Island, two miles south of the seaside resort of Tenby in Pembrokeshire, is home to one of the more extraordinary monasteries in the history of Christianity. There is a small village, but it is the monastery that dominates the island and the views from the mainland. The island is known locally as Ynys Pyr (the Island of Pyro), named after a sixth-century Celtic missionary. Its present name comes from the Old Norse *keld*, meaning spring, and *ey*, meaning island. The monks who inhabited it for many centuries, however, were not the first dwellers on the island. Human remains, flint tools and long-extinct animal bones have been found on the island dating back to roughly 8,000 BC.

The original monastery, built in the sixth century, was destroyed by Viking raids in the ninth century, but a Benedictine community returned to the island in 1136 and remained there for 400 years before it was dissolved, along with so many others, under Henry VIII. In 1906 an Anglican offshoot of the Benedictine order revived the community and in 1929 these monks were replaced by a Cistercian order who remain there to this day.

The sixteen monks on Caldey, who start their day at 3.15 a.m. with four hours of silence, aren't quite as unworldly as their medieval predecessors. Over the years this curious, near-silent community has had to display the guile of Brother Cadfael, the fictional medieval monastic sleuth, to stay in existence. In the early 1980s, the number of monks fell to just nine and there were fears that the community would no longer be able to keep going; but a worldwide appeal for recruits boosted its numbers, though many of the applicants, from as far afield as New Zealand, were turned away, some because they were not Roman Catholic. Money, or rather the lack of it, has been an ongoing problem for the community for decades, but rather than let the community fade away and die as so many others have, the monks determined to rescue their way of life by opening up their community to the very world from which they had originally sought isolation. Eight years ago the order came to national attention when it started advertising the monastery as a tourist attraction. Much of the money on which the monks and the forty or so lay residents of the island depend comes from the monks' own-label perfume, biscuits and chocolate.

I was intrigued by these entrepreneurial monks, who lived with one foot in the modern world and the other in the Middle Ages, and was keen to discover how they managed to pull off the seemingly impossible trick of living a monastic life while running a successful business. So I headed up the Pembrokeshire coast, stopping on the way in Dylan Thomas country to visit the great Welshman's writing shed in Laugharne. Laugharne is a small coastal village situated on the Taf estuary just a few miles from Pendine Sands. Thomas, who died in 1953 while on tour in the United States, is buried in the village's little churchyard of St Martin's, and his favourite watering hole Brown's Hotel, where he drank so lustily, is still open for business.

With its endless beaches and its striking craggy cliffs the Pembrokeshire coast must surely be one of the most beautiful in the British Isles. I had always been fascinated by Pendine Sands, best known as the location of numerous attempts to beat the world land speed record. In 1927 J. G. Parry Thomas was killed while trying to beat Malcolm Campbell's record of 175 miles per hour. His car, called Babs, overturned and he died instantly from horrendous injuries. He was still in the blazing car, partially decapitated and burned, when his crew arrived at the scene. In order to retrieve his body from the wreck his assistants had the dreadful task of breaking the legs off his corpse. The local people buried the remains of the car in the sand, but in 1969, amid much controversy, Babs was excavated and is now fully restored to her former glory.

Some friends I made during the year I spent on Taransay now live on Pendine and when I go to see them this stretch of coastline never fails to take my breath away. You wonder whether it is possible for nature to contrive a more beautiful coast. The great rugged cliffs towering over lovely golden sands are a little like parts of Devon and Cornwall, only bigger and wilder, and with nothing like as many tourists.

To get to Caldey I caught the small ferry that sails back and forth from Tenby, where Thomas delivered his first ever reading of *Under Milk Wood*, but which today is better known for its amusement arcades, ice-cream vans and beach shops. It was a beautiful spring day as my boat, packed with day-trippers, skimmed across the calm water from the mobile jetty on Tenby beach. Since the monks were forced to accept that farming alone would never sustain their small community, the island has become a huge tourist attraction, drawing around 65,000 visitors each year. Think monastery and I think peace, solitude and reflection, but there was little of those qualities in evidence as we pulled alongside the little jetty after the twenty-minute journey across the sound.

When I set out, I had been half-expecting to step straight into the set of Sean Connery's film *The Name of the Rose*, with a dozen or so men in brown robes, their faces obscured by hoods, their hands in their sleeves, walking slowly about the island in silent contemplation. Instead I found an island packed with pushchairs, buckets and spades, and sun-burned beer bellies, as hundreds of holidaymakers scoffed on local ice creams and chocolate, children leapt and squealed on the beach, and purple-rinsed grannies sipped cuppas in the tea gardens.

You cannot see this seaside resort life as you set off across the sound because your eyes are drawn inevitably towards the massive Caldey Abbey, which dominates the skyline. The present building was designed and constructed in 1907–15 by the architect John Coates Carter in Italianate style with red-tiled roofs and turrets. The Grade II Listed building is a striking sight, but my first impression was that it looked a little out of place on this windswept Welsh island. It did, however, look like a welcome refuge for the monks from the tourist mayhem that descends on their shores every day of the week except Sunday, when the island is closed.

As any owner of a large country house will tell you, running the pile can be a very expensive business and the extensive repairs the monastery needs have put a huge strain on the community's finances. In 1999, the National Lottery came to its aid when the Lottery Heritage Fund awarded a £400,000 grant for the repair of the roof of the monks' architecturally significant home. Previously, every time it rained they had had to put out buckets to catch water pouring through the roof.

It did not take me long to find my first monk. In the little shop near the harbour where the ferry had disgorged its cargo of noisy day-trippers, I came across Brother Robert O'Brien. I could hardly have missed him, to be honest, because with his long, bushy, white beard, bald pate and flowing brown robes he looked as if he had walked straight out of central casting for *The Name of the Rose.*

Brother Robert explained how a world away from the glamour of the big perfume houses, the monks produce scents with fragrance names that include Fern, Gorse, Brocade and Bouquet. One of their best-sellers is Caldey No. 1, not to be confused with the Chanel variety. All their perfume products are made from natural sources on the island. 'None of them, though, is called Temptation or anything suggestive like that,' he added. 'Some may think our scents quite old-fashioned, but I've been told that organic scents are actually coming back into fashion.' Self-sufficiency, even to live a life of austerity, which remains one of the main principles of this order, has come at a high price.

'Business was very good in the seventies and eighties,' explained Brother Robert, who originally comes from New Cross in south-east London, an environment about as far removed from an island monastery as it is possible to imagine. 'But in the nineties it began to erode quite badly.' The low-cost airlines and growth of local attractions in Tenby diverted people's attention. 'We realized that competition in the perfume market was great. I was amazed when once somebody took me to a big department store in London where the entire ground floor was given over to perfume sales. I couldn't believe it – there must have been hundreds of different varieties.'

They were obliged to expand their staple range of perfume and chocolate products and sell books, post-cards, poetry collections and CD recordings of plain-song. They run a café, which is supplied with milk, yoghurt and ice cream from their very own Jersey cows. 'We decided to see if we could sell more products off the island,' he explained. The monks now sell their goods around the world through the Internet. Should the bottom ever fall out of the world scent and choc-olate market, the monks could always turn to their organic meat enterprise: they produce prime steaks and joints from their thirty-strong herd of Hereford-cross-Friesian cattle, which graze on a 350-acre farmland rich in herbs. I made my own small contribution to the continuation of their way of life when I was relieved of £13.95 for a small bottle of Caldey for Men cologne, and a further £6 for what I must say were some very fine chocolates.

It must be a strange life for these monks, who seek a life of deep contemplation and try to live in near-silence but are forced to rely on the outside world, with all its materialism and consumerism, for their continuing existence. The irony is not lost on them, but it is admirable that they have accepted the reality of their plight and walked the path of compromise.

The modern world is camped outside their walls like an invading army, and occasionally succeeds in breaching their fragile defences. When Pope John Paul II died in 2005, the monks had to have a television specially installed at the monastery for the first time in order to watch his funeral. Because of the height of the building's walls, a satellite dish had to be rigged up to pick up the signal. But as soon as the broadcast of the service was finished, the dish and television were taken away.

The community found itself dragged into another arena of modern life when the monks were summoned to a tribunal in Carmarthen, accused of unfair dismissal by their former cook and seamstress. The story, predictably, was splashed all over the nation's newspapers. Brother Robert was allowed to speak at the tribunal, despite the order's vow to lead a near-silent life. 'It's not actually a strict vow of silence,' he told me. 'If it was I wouldn't be allowed to talk to you now, but it is a strict rule of the monastery not to speak unless necessary.'

Andrew McHardy and his wife Sally had lost their jobs of nearly twenty-three years amid accusations of heavy drinking and missing donations. During the hearing Brother Robert tried to explain that in an ideal world the community would be entirely self-sufficient but that they had to employ the McHardys because of a shortage of trained monks able to work. The hearing was told that the McHardys, whose duties included repairing habits, were made redundant when younger monks took over their duties to cut costs at the monastery. Ben Childs, the island's estate manager, described to the tribunal the precarious state of the community's finances, revealing that they had debts of £36,000 at the time he was taken on. The monks made a profit of £9,000 the following year and that fell to just £1,800 the year after on a turnover of roughly £350,000. Although the court heard that on asking the McHardys to leave, the monks had given them a modest 'friendship' payment, the tribunal found that they did not reflect the long service the McHardys had given them. The monks lost the case and were told to pay extra compensation to the couple. If nothing else, the case highlighted the ongoing difficulties the monastery faces in trying to sustain its ancient way of life.

As I walked around the beautiful grounds of the Abbey, I could not help but be affected by the enchantment of the place and its serene atmosphere. The Abbey Church was open to visitors for the monks' daily prayer and so I stole into a pew at the back and slipped off into a dreamy haze as the sixteen monks chanted their prayers. It was a world away from the string vests and the candy floss a few hundred yards down the hill. The church inside has a wonderful simplicity to it, which makes a perfect environment for prayer and reflection, belying its grandiose, ornate exterior. Outside, the cloister, set around a central garden, provides another silent sanctuary for quiet contemplation. A short walk away is the refectory, an imposing oak-panelled hall with a timbered roof, where the monks take their communal meals.

The monks' first vigil of the day at 3.15 a.m. is followed by private prayer, meditation or study until 6.30 a.m., when it is time for Laud and Concelebrated Mass. Terce takes place at 8.50 a.m., before the monks begin work or study. At 12.15 p.m. they come back to the Abbey Church for Sext, before returning to their labours. At 5.30 p.m. it is Vespers, followed by Compline at 7.30 p.m. The monks then retire at about 8.00 p.m. and observe a strict rule of silence until seven the following morning. What an odd life it must be for them, I thought, dashing from the peace of the cloister and the Abbey Church to stand behind a counter dispensing ice creams and confectionery while being gawped at and photographed by boatloads of day-trippers!

The tourists rarely seem to venture much beyond the little harbour area and the beach, and there is plenty of solitude and natural beauty to be enjoyed on the rest of the island. There were only a handful of 'outsiders' to be seen when I took myself off for a walk along the path that circles the island, which is about one and a half miles long and three quarters of a mile wide. There was, I discovered, far more to the island than just the monastery. There is the Old Priory, one-time home to the Benedictine monks in medieval times; built on one of the island's highest points, close to a natural spring, it stands derelict next to the farmyard the monks work today. The neighbouring St Illtyd's, originally the Priory Church, is still a consecrated Roman Catholic church. With its leaning spires and pebble floor worn smooth by generations of monks and worshippers over the centuries, the church retains a very strong spiritual atmosphere.

In the graveyard of the parish church of St David,

there are rows of old wooden crosses marking the graves of monks and islanders. Historians believe the church stands on the site of a pre-Christian Celtic burial ground, dating back about two thousand years. It is possible that the site was used for burying people from the mainland, in accordance with the Celtic belief that islands represent a bridge between heaven and earth.

As I walked up to the lighthouse, I was met by stunning panoramic views of the Pembrokeshire coast, the Preseli Hills, the Gower peninsula and Lundy Island. Built in 1829 to aid navigation in the northern Bristol Channel, Caldey lighthouse was originally powered by oil and required resident keepers to keep it in operation; it was not until 1997 that it was fully modernized and converted to mains electricity. As I stood there with a firm breeze blowing off the sea and a glorious sun beating down on the fields and rocks around me, it was not difficult to understand why pagan priests, monks and worshippers have kept returning to this holy island for at least two thousand years. Nor was it difficult to understand why the present inhabitants have been hanging on to their way of life so tenaciously, even if that means dishing out ice creams to godless day-trippers in string vests.

Much as I admired the Caldey community for their resourcefulness and their serenity, I had to face up to the cold fact that I myself would make a dreadful monk. 'Brother Ben' just doesn't have a convincing ring to it. Apart from the fact that I have simply no idea what words like Lectio or Terce mean, and apart from the fact that I am a confirmed agnostic, it is the vows of poverty, austerity, chastity and obedience that would soon find me out. Getting up at 3.15 a.m. every day and giving up sex? Not for all the chocolate on Caldey Island.

## Bardsey

If you were to sail west from Caldey, take a sharp right at the end of the Pembroke peninsula and head north through Cardigan Bay, you would eventually come to a tiny outcrop off the Llyn peninsula, which, like its southerly cousin, has also been struggling to hold its own against the powerful tide of modern life.

Bardsey Island is a tiny stretch of land, less than two miles by one, famous for its birds, its plants, its peculiar history and its place in the Arthurian legend. Unlike Caldey, its problems are more geographical than religious. Cut off from the mainland for days and even weeks at a time by strong tidal currents and battered by the Irish Sea, Bardsey has a recent history of gradual depopulation and a struggle to reverse the trend.

At the beginning of the twentieth century, the population of the island was around a hundred. In 1964, the great travel writer Eric Newby visited Bardsey, which he dubbed the 'disenchanted isle', to discover a population of just seven. He speculated that the number would dwindle still further, and he was proved right.

By 2000 the population had dropped to just three – two nuns and a farmer, who were, let's face it, hardly likely to trigger a baby boom on the island.

The Bardsey Island Trust, which has managed the island since buying it in 1979, decided to take matters into its own hands. 'Professional Castaways Wanted!' announced advertisements in the national newspapers. It might have seemed like a last throw of the dice by the custodians of this beautiful and historic island, but was there much difference, I wondered, between this ad campaign and those of cities and big towns that appeal from billboards and magazine pages for people to move there? Certainly, the stakes were much higher, but advertising struck me as a perfectly sensible course of action, given the island's desperate plight, and given the fact that the vast majority of the British public have probably never even heard of Bardsey. If the trust could persuade even a tiny fraction of them, perhaps a dozen or so families to come and live in Bardsey, the future of the island would be guaranteed, at least in the short to medium term. Having spent a year as a castaway myself, I was highly intrigued to see how successful the island had been in attracting permanent settlers.

Bardsey Island is like a comma punctuating the Welsh coastline. Its Welsh name, Ynys Enlli, means 'island in the current', which refers to the surges of water that funnel through the strait, making access unpredictable at best and extremely hazardous at worst. The three different currents that swirl around the island have posed a mighty hurdle for visitors over the centuries – and that very inaccessibility, from which arises the fear of being marooned, is central to its problems today.

The island can be reached in the summer by ferry from the popular seaside resort of Abersoch, a few miles away on the Gwynned coast: it departs every few days or by prior arrangement with the skipper. I found myself stormbound in Abersoch for three days, stewing in a small B&B while I waited for a break in the weather. I whiled away the time sipping tea, reading up on Bardsey's extraordinary history and staring out at the rain hammering on my bedroom window – which perhaps provided the answer to why Bardsey found it so difficult to attract settlers: Wales, it must be agreed, can be a particularly dour place in the rain.

Finally the wind and rain subsided and gave way to a warming sun, and I wasted no time in racing down to the tiny harbour to secure my short passage to Bardsey. We passed along the dramatic Llyn peninsula with its steep cliffs and in the distance I could make out the icing-sugared pates of the Snowdonian mountains; and as we cut through the relatively calm waters we passed 'porpoising' seals and dolphins diving in and out of our bow wave. When Bardsey loomed into view I couldn't help but wonder how tens of thousands of pilgrims managed to reach this remote place, where nature does its best to deny access to outsiders. Even with the aid of a modern fishing boat we battled against the powerful currents for more than two hours before reaching it. As we drew closer, I was amazed to see a giant swirling cloud of birds swooping and diving over one end of the island. These, the skippers told me, were the famous Manx shearwaters, about twenty thousand of them, who colonize the island.

Bardsey, I could see, is an island of two distinct halves, with a sharp ridge on the north-east dissolving into lowland to the south and west. A lighthouse tops a steep hill at the end of a peninsula. The views around it are truly spectacular in every direction, with the great watery expanse of Cardigan Bay to the east, the pale silhouette of Anglesey to the north and the hills of the mainland marching away from the coastline eastwards to the cloud-draped peaks of Snowdonia.

There are dozens of myths and legends about Bardsey but remarkably few hard historical facts. Incredibly, it was not until the 1960s that the first serious archaeological investigation took place. Bardsey has been noted as a place of pilgrimage since the early years of Christianity, but there are signs of settlements on the island that date from earlier periods. A great number of earth structures have been identified, some of which date back almost three thousand years. The island became a focus for the Celtic Christian church and it is believed that St Cadfan, a Breton monk, began building a monastery on it in the sixth century, at roughly the same time that the monks of Caldey were constructing theirs down the coast. Since the Middle Ages there have been references to Bardsey as the burial place of 20,000 saints, prompting one of the Popes to decree that three pilgrimages to the island equalled one visit to Rome. With Bardsey, it is impossible to say where the facts end and fiction begins.

One fact of which I could be sure was that I was to spend a night on this mysterious island in an abandoned farmhouse with no running water and no electricity (there are also no cars on the island). I couldn't help but wonder, after my Abersoch experience, how I would feel if a storm came in off the Atlantic and left me marooned here for days, perhaps even weeks. I just hoped I liked it there – and that the inhabitants – there were now five of them – liked me.

We inched our way carefully into the harbour, if that's the right word for an incredibly narrow inlet blasted out of the rock, and I could see a single dirt track winding its way into the island's hinterland. I was the only visitor to Bardsey that day and I cut a lonely figure as I stepped ashore and the skipper of my little boat set about unloading the shopping delivery from the mainland stores. Simon Glynn, the director of the Bardsey Island Trust, was at the rocky landing point to greet me. 'Welcome to Ynys Enlli!' he said, beaming.

It was Simon who had masterminded the island's national ad campaign for 'professional castaways'. I soon discovered that it had yielded surprisingly few serious enquiries. Even the offer of one of the very presentable 130-year-old three-bedroom farmhouses to live in, with a school for the children thrown in for good measure, had failed to persuade potential candidates to part with the estimated £25–30,000 they would be expected to invest in the island. (It wasn't like that for Robinson Crusoe.) Interested parties had been told that the money was needed to invest in a flock of sheep, a tractor and a fishing boat for travelling to the mainland. In return, settlers would have a twenty-year lease on their property and the promise of a good life in undeniably stunning surroundings.

Could Bardsey be the dream island I had sought for so long? I cut to the chase. 'If I took the place and met your demands, would you be prepared to declare me King?' I enquired.

'Actually, we already have one, I'm afraid,' Simon informed me proudly. 'King Bryn Terfel.'

The story of Bryn's coronation is as colourful and surreal as it is controversial. In 1999, the Bardsey Island Trust offered the Welsh baritone the Bardsey throne, which had been vacant since the death in 1927 of King Love Pritchard, a rather eccentric character to say the least, who had been declared King by Lord Newborough, the former owner of the island. When King Love made a rare visit to the mainland, he was welcomed by Prime Minister, Lloyd George, as an 'overseas King', and the *Daily Sketch* ran the headline 'King Love Arrives on Mainland.'

The throne lay unclaimed for over seventy years until the Trust decided to offer the crown, a modest affair of brass and tin, to the famous Welsh opera singer. That, however, stirred up a constitutional hornets' nest, as a man called Ken Pritchard, a middle-aged furniture salesman from the Midlands, claimed that the throne was rightfully his, as he was the great-grandson of King Love Pritchard. 'My grandfather was an only son, as was my father, and so am I,' Ken told the press. Ken said he had no desire to be crowned King and was simply speaking out to claim his family's birthright. The crown is stored safely in the Liverpool Maritime Museum and the family are campaigning to keep it there.

'We were inundated with people claiming to be descendants of King Love,' Simon said. 'I'm sure most of them are very sincere people, but until we can research each claim, King Bryn Terfel remains the monarch.'

Bryn Terfel is certainly not the first royal ruler of an eccentric offshore kingdom in British and Irish waters. If you travel to Tory Island, a remote, picturesque island nine miles off the coast of Donegal, you will meet Patsy Dan Rogers, artist, musician and king, whom I have had the great honour and pleasure of meeting. From the earliest times, the inhabitants of Tory had seen themselves as a different people from the Irish across the water and clung to their ancient ways. The current social organization on Tory dates back to the sixth century, and Patsy followed a long line of island monarchs. The title of King of Tory, however, is not hereditary and kings are often elected for their skills and personal qualities. When the last king died, his son turned the job down on the grounds that it was too much work and responsibility. Patsy landed the job because no one else on the island could be bothered.

Lundy, an island of great beauty in the Bristol Channel, also has a history of a peculiar monarchy. A former home to French pirates, Lundy passed from aristocratic ownership in the nineteenth century when it became the property of a character called Martin Coles Harman, who in 1925 duly declared himself King and set about issuing stamps and Lundy currency in the form of half puffin and one puffin coins. Residents did not pay taxes to England and visitors had to pass through customs when they travelled to and from the island. The House of Lords, however, took a dim view of King Harman's actions and charged him with violating England's 1870 Coinage Act. The King was summoned to court, where he was duly fined £10. The coins were withdrawn and became collectors' items worth a small fortune today.

These tales continue to give me hope that one day I too will be crowned king of my own proud little island, but Bardsey, it seemed, was a constitutional mess that I would do best to avoid. The rebuff still ringing in my ears, I decided to leave Simon and explore the island on my own. I ambled along the narrow road – the only road on the island – and past the working farm, run by Colin Evans, who doubles as the island's lobster fisherman.

The views of land and sea on Bardsey were breath-taking as I made my way to the ruins of St Mary's Abbey. It's no coincidence, I thought, that centres of pilgrimage and spiritual retreat such as Caldey and Bardsey tend to be places of outstanding natural beauty. Bardsey is a National Nature Reserve and a Site of Special Scientific Interest and, despite its size and lack of inhabitants, you could spend days exploring and enjoying its charms.

The wildlife attracts visitors from all over the world who, generally in the summer months, come in their droves to admire the birds, rare flowering plants, lichens, mosses, coastal grassland and heathland, and marine wildlife. (Botanists may be interested to know that on Bardsey there are 350 species of lichen.) Atlantic grey seals play in the rocky bays of the island, while dolphins and harbour porpoises are a common sight in the waters further out. The island is a twitchers' paradise, lying as it does in the spring and autumn migration paths of many birds. It is home to choughs and oystercatchers and you also see peregrine falcons, warblers, herons, owls, gannets, razorbills and shags, as well as the huge colony of Manx shearwaters I witnessed on my arrival. The waters around the island, with their forests of strap seaweed, are rich in marine life. In the rock pools you can see anemones, crabs and small fish, and in the deeper waters filter-feeders such as sponges and sea squirts carpet the rocks.

What's peculiar about Bardsey is that most of the houses you come across are genuinely attractive, roomy and architecturally impressive. They almost look a little out of place, as if they would be better suited to a street in the smart area of a northern town. Most houses on our islands are low squat affairs with poky windows, designed and built not to please the eye but to keep out the weather. Most of the Bardsey houses were built by Lord Newborough in the 1870s and although they are all virtually empty, but for a few camp beds and the odd table, they are in remarkably good condition and could quickly be transformed into comfortable family homes. If they were located in southern England, they would probably fetch around half a million pounds each.

I reached a cluster of abandoned farmhouses and cottages, about a dozen in total, that stand mournfully in the north of the island, around the ghostly silhouette of St Mary's Abbey, which was built on the ruins of the original monastery. It was strange to reflect that for many centuries this site was the home to thriving communities, who fished and farmed for a living. Even at the start of the twentieth century around a hundred people lived here, but the majority of these followed King Love off the island in 1925 in search of a less laborious way of life.

'Respect the souls of the 20,000 saints whose remains lie close to here' instructs a memorial stone in the graveyard. Many of the pilgrims to Bardsey came to die. Many others perished on the treacherous journey. The thirteenth-century Augustinian Abbey of St Mary was in use until the dissolution of the monasteries in the 1530s, after which the island was abandoned to pirates, marauders and other ne'er-do-wells until people began to return to re-establish the community in the 1700s.

I paid a visit to the island's wardens, David and

Patricia Jones, who lived in a small cottage at the top of the road. The couple, who made up precisely two-fifths of the island's population, had been on Bardsey for two years and were responsible for maintaining the houses and looking after visitors. They were on their hands and knees busily planting potatoes in the rich soil, looking a little like Tom and Barbara from *The Good Life* in their thick jumpers and grubby overalls. Smiles radiated from their gently weathered faces as they happily explained how they were often stranded for up to four or five weeks at a time. 'We like the isolation,' said David, grinning and wiping a muddy sleeve across his sweaty forehead. 'We just make sure we are always well stocked,' added Patricia.

I had been assigned one of the houses that had been built by Lord Newborough. The three-bedroom house, called Plas Bach, was largely void of furnishings. There was a single sofa in the front room and a large pile of books mainly of an ornithological nature. The bedrooms had only camp beds in them and the kitchen consisted of a bottle of gas connected to a stove. A well outside provided water, while answering the call of nature involved a windy journey across the garden to a low shed with a saloon door.

Visitors to Bardsey have to bring all their own food and drink, as the nearest shops are two hours away – or five weeks away if you're unlucky – in Abersoch. As a tribute to my student days I had stocked up on 'boys' food', which consisted mainly of Doll noodles, tinned soup and a couple of bottles of beer.

Incredibly this pleasant but otherwise unremarkable house is said to have a significant place in history, or at least in legend, as the home of 'King Arthur's apple'. Bardsey hit the headlines just a few years ago with the discovery that the apples growing up the side of the house were one of the oldest known varieties in the world, dating back over a thousand years. The gnarled and knotty old tree with its pinkish lemon-scented fruits is the only survivor from an orchard tended by monks in pre-medieval times. It seemed amazing that these twisted branches had survived centuries of storms and neglect, and might once have borne fruit for King Arthur, Merlin the Magician and other heroes of the Arthurian legend: many have argued that Bardsey is the mythical island of Avalon, which means 'place of apples'.

The discovery of the tree's ancient origins came after a chance conversation between a birdwatcher who used pieces of apple to lure birds and a visitor who happened to be a fruit tree expert. He returned to England with some samples for experts from Brogdale Agricultural Trust in Kent to identify. They established the apple's antiquity and christened it the 'Bardsey'. I found it exhilarating to think that I was standing on the very soil that might have been the island where King Arthur came after the battle of Camlann so that his wounds could heal. Merlin is said to have been put to rest here in a glass coffin on the hill a few hundred yards away. Legend has it that Arthur planted many apple trees on the island, but the 'Bardsey' was the only one that would grow in the harsh conditions. Curiously, the apple is completely disease resistant, which is unique for fruit in north Wales.

The mother tree bore a surprising number of boldly striped 'pink over cream' apples and, after looking around to ensure that no one was watching, I plucked one. It was still early in the season and the apple wasn't yet ripe, but it was perfectly formed and looked delicious. I rolled the apple around in my palm, and held it close to my eye. It seemed extraordinary that this small apple could hold such an important place in history, in such an impossibly unlikely setting, stuck as it was in a slight recess on the side of a farmhouse, lashed flat against the wall by years of salty gales on a remote, virtually uninhabited Welsh island. But then with island life I had come to expect the unexpected. I was staggered to see that the old tree was unprotected and unfenced.

I took the small, sweet-smelling apple and bit into its tough skin. I was eating history. It was odd to think that King Arthur might have experienced the very same bitter-sweet taste, standing on this very spot. The island he would have looked out on had probably changed very little over the years, apart from the dozen or so houses that had been built.

The wind had disappeared and walking around this near-deserted 'ghost' island felt otherworldly. It was eerily silent and the moon cast shadows across the valley. With no light pollution, the island had an ethereal feel. I wandered across the barren fields and along the tracks, my way illuminated by the light of the full moon, which was so bright that I could make out the colours around me. The only noise was the occasional crash of water as a grey seal broke the surface of the calm waters.

I dropped in to see Colin, the farmer and lobsterman, and, over a bottle of beer, he told me how it felt to live on Bardsey. 'It's good for the soul,' he said. 'I have a sense of belonging.' That sense was one I had often heard expressed by islanders around the world. I wondered whether I could ever find an island on which I felt I truly belonged. I had been moved by my island pilgrimage. I felt sad to leave Bardsey; I had felt calm and relaxed while I was there, after those three frustrating days kicking my heels and pacing up and down my B&B room in Abersoch. Not normally affected by spirituality, I had been touched by my stay on the holy island.

I sat at the helm of the little fishing boat, contemplating what it would be like to live on this remote Welsh island, which is cut off so often and for so long. The stay had definitely been a highlight of my island odyssey, but I don't think I could ever live there. Bardsey made me realize that my perfect island would have to have a larger community. Until Simon manages to persuade more people to come and live there, Bardsey will always feel a lonely place. Bardsey was full of ghosts, not people; I couldn't help but feel that this was an island with more of a past than a future. But while it might not have been the island home I had been searching for all these years, I knew a couple for whom it might be perfect. Patrick and Gwyneth Murphy, a former postman and dinner lady from Preston, had been searching for such a place to retreat to. I had spent a year living with them on Taransay in the Outer Hebrides. They had thrived there, becoming the 'rocks' of the community and adapting readily to the cruel climate. They had returned to city life but had been unable to settle again, and had set their sights on returning to the island life they had enjoyed so much.

'I've found you an island,' I hollered excitedly down the phone, following my return to the Welsh mainland. 'How soon can you get here?'

'Tomorrow!' came the reply.

The next day Simon joined us and we all sat huddled in a little greasy spoon café in Abersoch, as rain streamed down the misted windows. Simon explained the responsibilities that would come with the island tenancy and it was agreed that Patrick and Gwyneth could have a trial period to give them time to decide whether to make a longer-term commitment.

David and Patricia, the island's current wardens, had expressed a wish to move on, and Simon explained that should the Murphys take to life on Bardsey, the job would be theirs. As I sat there, listening to them map out plans for their future, I was struck by the similarity between Patrick and Gwyneth and the Joneses, not just in their mannerisms but in their physical appearance. It was uncanny.

So that was simple enough: over a cup of tea and some laver bread, Patrick and Gwyneth had a new island home and King Terfel might have some new subjects. Bardsey was not for me, but at least I left in the satisfying knowledge that I had helped out some old friends and made a modest contribution towards revitalizing the community on that ancient, magical island.

## The Skerries

'*Mayday! Mayday! Mayday!*' the voice screeched over my crackling radio. 'The helicopter's ditched!' With that it went dead and I found myself marooned on a rocky island being lashed by a great storm rolling in off the Irish Sea. I was a castaway on one of the Skerries, some wild islets two miles off Carmel Head on the north-west corner of Anglesey, North Wales, cut off from the mainland by some of the most ferocious seas around our shores. I felt a sense of panic and nausea rising in me as I contemplated the fate of the two pilots who had just dropped me on this uninhabited outcrop.

I had truly landed myself in it this time, I thought, as I took shelter from the howling wind behind a large rock. The Skerries had never been part of my original itinerary. I had ended up there because a few months earlier I had applied for a booking aboard THV *Patricia*, the flagship of the Trinity House fleet, which is responsible for the maintenance of buoys and light-houses around our shores as well as dealing with emergencies and wrecks. The 2,400-ton vessel also happens to double as a luxury passenger ship. Twenty years ago, six cabins were converted to form a small de luxe floating hotel for the intriguingly named Elder Brethren, Trinity House's board of governors. They were fitted out with comforts that would have put many commercial cruise liners to shame. All the cabins are en suite and come with satellite television and telephones. The cheapest cabin, a single 'executive' room, will set you back just under £2,000 for a week, while the privilege of staying in the state room (and sleeping in the bed that the Duke of Edinburgh occasionally occupies) will relieve you of a sum closer to £4,000.

The *Patricia*'s itinerary is dictated by need rather than want and I had no idea where I would be heading when I finally received my call from Trinity House. When I was contacted I was told I had two days to get to Holyhead in Anglesey, where I would join the ship before it continued its way to the Skerries. I had never heard of the Skerries, but some hasty research informed me that they lie in the middle of an area of shallow rocky outcrops and vicious ocean currents. These windswept, treeless islands are inhabited only by seals and birds and are home to just one building, a lighthouse. The origin of the name is Norse, a *sker* being a rocky islet. The Welsh call them Ynysoedd y Moelrhoniaid, which means island of bald-headed grey seals.

A light had been proposed on the Skerries as early as 1658 by a private speculator, Henry Mascard, who hoped for substantial profits from tolls levied on passing ships. This corner of Anglesey was notoriously treacherous and mariners were eager for a marker there, but nothing came of Mascard's proposal because of the opposition of Trinity House, the national body established in the previous century for the purpose of aiding navigation in British waters. The dangerous rocks continued to claim their victims, including the former royal yacht, HMY *Mary*, which sank in 1675. The *Mary* was built by the Dutch East India Company and given to Charles II on his restoration to the throne. The great diarist Samuel Pepys sailed in her, and she was used for official journeys and royal leisure trips before her unfortunate demise on the Skerries. She has since been partially excavated and some artefacts from the wreck can be found in Liverpool Museum.

In 1714 Queen Anne granted William Trench permission to build a light on the Skerries and levy dues. He was permitted to collect a compulsory fee from all passing shipping for the upkeep of the light, in return for paying an annual rent of £5 to the Treasury. This potentially lucrative venture turned sour for Trench. During the construction of the light he lost his only son in a shipwreck and he had great difficulty collecting his dues. The light was first kindled on 4 November 1717, but Trench fell into heavy debt and he died a ruined man in 1725. The lease and the right to maintain the light passed to his wife, who conveyed them in turn to her daughter and son-in-law, the Reverend Sutton Morgan, who initially seems to have fared no better than his father-in-law.

However, the light had proved its worth and to guarantee its future, Parliament passed an Act in 1730 that assigned the Skerries to Morgan and his heirs in perpetuity, at the same time raising the tolls due from passing ships. The Act caused great embarrassment to Trinity House, as over the next century or so the Skerries became the most profitable lighthouse in England and Wales, thanks to the great volume of traffic heading in and out of the thriving ports of Liverpool and Holyhead. Trinity became increasingly desperate to take it under its authority, but when a new Act was passed in 1836 giving Trinity the power to acquire coastal lighthouses, the family refused to hand over its business until finally a court awarded them a staggering £444,984 (the equivalent of £22 million in today's money) in compensation. The Skerries lighthouse was the last privately owned light in the British Isles to be purchased by Trinity House and by far the most expensive. When Trinity House took it over it halved the light dues but nevertheless continued to make a huge profit.

Trinity House still owns most of the lighthouses around Britain, though the departure of the last resident lighthouse keepers from North Foreland lighthouse in 1998 marked a significant change from traditional working practices. Automation transferred most of the work to Trinity House's headquarters in

Harwich. Today the keepers have gone, swept aside by technology and replaced by a number of shorebound watch keepers who keep an eye not on a vast ocean but on a switchboard. While the rocks and skerries have passed back to the razorbills and the guillemots, the puffins and the gulls, technical staff must still carry out preventative maintenance of the lights: there are windows to be cleaned, bulbs to be changed and oil to be replenished. These responsibilities fall to the *Patricia*.

I duly made my way to Holyhead, where a bright orange tender came to the docks to ferry me and a dozen or so other passengers out to the *Patricia*. The sky, streaked with various shades of black and grey, looked menacing as we climbed aboard the small craft. There was a strong wind up and even in the sheltered harbour there were plenty of white-topped waves that sent spray crashing over the front of the tender. The crew pulled up a tarpaulin to protect us from the onslaught of the sea. On board the *Patricia* a party of officers greeted us each individually, perhaps full of admiration for a group of people willing to spend the better part of £2,000 each for a week in stormy waters watching engineers replace light bulbs and clean windows. We were escorted to a smart passenger lounge full of leather chairs and sofas with a large fireplace in the middle and a portrait of Her Majesty The Queen on the wall. The bookshelves were lined with maritime novels and board games, a sure sign that *Patricia* was well prepared for many long wet days at sea.

Since HMY *Britannia*, the royal family's luxury water palace, was decommissioned, the *Patricia* has taken on the role as the family residence at sea and Prince Phillip, who is also, incidentally, the Master of the Elder Brethren, regularly uses the ship. The crew keep a stock of light ale for him, and a bottle of Dubonnet in case the Queen should ever decide to join her husband for a spot of lighthouse-gazing.

Before *Britannia* was taken out of commission, it was the prescriptive right of the Elder Brethren to escort Her Majesty The Queen to sea in home waters. As the flagship of the Trinity House fleet, it fell to the *Patricia* to guide the royal yacht out of Portsmouth, seeing the *Britannia* clear of the dangers of land before detaching and executing a smart 'steam past', with her crew at the rail and the Elder Brethren paraded on the helicopter flight deck. This little ceremony was invariably acknowledged by Her Majesty in person.

*Patricia* is fitted with special towing winches, sufficient to pull a fairly large ship away from a dangerous situation as well as providing a routine capability for moving light ships to and from their stations. Cranes point skywards from either end of the vessel and navigation buoys line her decks in readiness for laying, but despite being a working vessel operating in some of the harshest waters around our coast, the *Patricia* has an air of grace about her. She is certainly striking on the eye with her gleaming black hull and a helicopter secured on her stern. She dwarfed every other boat in her mooring at Holyhead.

The other passengers on the ship included a doctor, a dentist, two schoolteachers and a posse of former naval and merchant officers, their shoes still polished to within an inch of their lives. We all sat in the lounge as the rain beat against the panes and crew scurried around securing the ship for departure. My coffee cup slid up and down the table with each lurch and roll of the ship.

'Welcome aboard,' announced the captain. 'I'm afraid we're in for some bumpy weather today.'

'Splendid!' beamed Charles with genuine delight. 'Splendid!' Charles, needless to say, perhaps, was a former naval commander, now in his sixties, who loved nothing more than a jolly good storm. It turned out that this would be his third voyage aboard the *Patricia*.

The captain explained that we were on our way to carry out some emergency repairs to the Skerries lighthouse, which had been damaged in a storm. 'We have an easterly force eight, increasing to nine, so I would advise you all to take a seasickness tablet.' My stomach heaved at the thought of the journey ahead. I was already feeling queasy and we hadn't even left the shelter of the harbour. I took four pills, fearing the worst. The enormous anchor was then winched aboard, the clanking of her heavy chain almost drowned by the screeching wind and crashing coffee cups.

Despite many changes over the years, buoy maintenance remains very much the bread and butter of *Patricia*'s work. Work is undertaken in surprisingly poor conditions and occasionally, when a new wreck requires urgent buoying to warn shipping of its un-charted position, buoys are laid in extreme weather. The *Patricia* is fitted with a 20-ton speed crane, capable of lifting the largest navigational buoys. It is not unknown for a 10-ton buoy, with a 5-ton cast-iron sinker and 328 feet of heavy steel chain to be laid in wind strengths of force nine.

It wasn't always like this. In the days of oil-gas and acetylene buoys, the occasional extinguishing of a lantern in extreme weather required the prompt attention of a Trinity House vessel. The motor launch was often lowered in appalling storm-force conditions, and the coxswain would bring it as close to the bucking buoy as he dared. An officer and a seaman would then jump on to the defective buoy and relight it, after which came the really difficult job of getting off it without injury. This process would often take hours and some fine judgement on the part of the crew. Today the old-fashioned jump in such circumstances has been replaced by that most modern of work-horses – the helicopter, a Bolko 105, which is used to transport personnel and material back and forth. Two pilots escort each voyage, taxiing crew from ship to lighthouse.

The passenger lounge overlooked the ship's small helipad. The leather chairs were all arranged so that passengers could keep an eye on their Patrick O'Brian novels and one on the helicopter. The passengers watched and scrutinized every move of the crew as if

they were Darcey Bussell and her troupe of baller-
inas with the helipad their stage. Occasionally heads
were shaken in indignation and the lounge was filled
with the sound of 'tuts' and grumbled mutterings:
'Where's his hard hat? . . .' 'I would never have done
it like that . . .' Old habits. The crew of the *Patricia*
have grown accustomed to lectures or advice from
the ex-captains and officers who often make up the
passengers.

The ship bucked and jerked its way across the heavy
ocean. 'Healthy force eight, this one,' observed Charles,
his 'bins' trained on the horizon. 'Should get a bit
sloppy soon.' He smiled indulgently. Out on deck, the
crew scurried around in luminous orange jumpsuits,
preparing materials for transfer to the island. The helicopter, heavily harnessed to the deck with ropes and
ties, pulled and strained with each lurch. Scrabble tiles
jumped from the board and cups leapt from saucers
with each fall as the ship pounded into the waves. I
noticed that our party had been significantly reduced
as the storm grew in strength and green faces retired
to their cabins. Soon, the Skerries appeared shrouded
in a mist of white spray and foam as vast waves crashed
against the shoreline. It seemed impossible that anything could fly in this.

About a mile from the lighthouse rock, the captain
turned the ship side on. The new motion was even
more uncomfortable: the ship rolled from side to
side with a stomach-churning inconsistency. It was,
however, a sensation I wouldn't experience for much

longer, as I had been offered the spare seat in the chopper. 'You'll have to watch the safety video first,' announced the pilot, leading me into the bowels of the ship to a windowless wardroom. I sat there watching the screen, feeling increasingly nauseous as the ship rocked and twisted this way and that. I began to sweat and I could feel the colour draining from my cheeks as I festered in the stuffy room.

To make matters worse, I was asked to don one of the lurid orange suits worn by the crew. Made of rubber, these sea survival suits can be insufferably hot and restrictive with their tight seals that garrotte the neck. How did I come to find myself in this bizarre situation, I wondered, as I clambered aboard the helicopter and stared out at the raging sea beyond? The chopper rocked and rolled with the ship's movements and salty spray streamed down the bulbous windows. I watched as the pilot flicked and turned and twiddled and spun switch after switch. Why is it that helicopter pilots always have an air of masterful control, even arrogant superiority, about them? Are they all six foot six with chiselled features and well-set jaws? Maybe they all come from the same sperm bank, I thought, pondering.

The chopper shuddered violently as the blades rotated and the helicopter lifted away from the deck. The wind whipped and lashed against us, and below, the sea had turned into a bubbling cauldron of white caps, foam streaming across the surface. Soon we were over the rocky plateau of the island and going towards a small, improvised helicopter landing area. 'It may get a little bumpy,' the pilot said, grinning devilishly as a great gust of wind funnelled off the rocks and threw us around violently. I clutched the seat, a bead of sweat chased down my brow. And then with a thud we were down.

The pilot signalled for me to get out and before I knew what had happened I was scurrying like a crab as the chopper disappeared into the murky horizon. Visibility had decreased and the ship had disappeared into the squally mist. Suddenly I felt very alone, as the wind whipped at my big orange Babygro and I headed off in the direction of the lighthouse, the only form of shelter on this wretched, storm-blasted islet.

In the pre-automated days the lighthouse keepers were isolated here for months at a time. I had been on the island for barely five minutes and already I was looking forward to leaving. The horrendous weather was the main problem, but even on a glorious summer's day I couldn't imagine that there was much to explore. There appeared to be nothing of great natural or historical interest. This Skerries island was a very bleak place indeed, and I wondered how those lighthouse men of old were able to stand the loneliness. Perhaps they had a tremendous supply of books or were born with the gift of enjoying their own company for days on end.

Watches were divided into day and night. In the dark hours, maintaining the light and ensuring that it never went out involved considerable effort. As late as

Caldey village, dominated by the island's monastery (© *Caldey Island Estate/Mimie O. Brien*)

t David's Church and graveyard, which is believed to stand on a pre-Christian Celtic burial ground
(© *Caldey Island Estate/Mimie O. Brien*)

A helicopter comes in to land on the deck of the *Patricia*, to take us to the Skerries

The wild and desolate Skerries …

... and their empty lighthouse

The small harbour of Creux on the east coast of Sark (*Robert Smith/Chapel Studios*)

La Coupée, one of the most spectacular sights in the Channel Islands: a 300-foot-high narrow isthmus separating Little Sark from the main island (*Robert Smith/Chapel Studios*)

Beautiful rustic scenes from the Isle of Man

Eigg's Singing Sands (*courtesy of the Isle of Eigg Heritage Trust*)

The imposing An Sgurr dominates Eigg's landscape (*courtesy of the Isle of Eigg Heritage Trust*)

Approaching the dramatic island of St Kilda

Reminiscing with Norman outside his old village on St Kilda

*(Above)* The Outer Hebrides,
the starting point for our trip to St Kilda

*(Left)* St Kilda's stunning coastline

the 1970s, heavily weighted clockwork mechanisms had to be wound up at regular intervals of twenty minutes. Similarly, until the advent of the compressed-air-driven foghorn, warnings involved firing hand-held explosives, which was a laborious task in the chilly, fog-shrouded lantern gallery. It was not unusual for keepers on the Skerries to be stormbound. This must have been morale-sapping, particularly when, as often was the case, it happened at Christmas. Depression was common among lighthouse keepers.

As I scrambled up the rocks, the hand-held VHF radio I had been given crackled into life and it was then that I heard those dreaded words cry out: 'Mayday! Mayday! Mayday! The helicopter's ditched!' My jaw dropped and my heart sank. I felt physically sick as I tried to take on board what had happened to the chopper in which I had been sitting just moments earlier. What did the voice mean by 'ditched'? Had the helicopter crashed? Were the pilots OK? Was it in the ocean? Or had it crashed into the ship? What if the ship was on fire? I was overwhelmed by un-answered questions. The radio went dead. I was marooned. The cloud had descended and the ship was just a faint blot in the distance. What was I supposed to do now? I was stuck on a remote island in a force nine. My helicopter had crashed and I had no idea whether the crew were dead or alive, and there was no way of getting back to the ship. I had no food.

I walked around the empty lighthouse, trying to work out the contingencies. A thick mist had descended over the island and her foghorn had been automatically activated; it eerily reverberated around the island as I wandered in and out of the now derelict keepers' cottages and along the rusty railway tracks that had been laid during the construction of the red and white tower. I climbed the spiral staircase to the glass top of the tower where an enormous lantern rotated on a bed of mercury. A tiny steel door led to a narrow parapet that surrounded the tower. I forced the door open against the powerful wind and climbed out on to the vertiginous balcony. The wind was like an invisible wall. I could lean into it and relax against its brute force. I could just make out the *Patricia*, her hull rising and falling in the rough seas. There was no sign of the helicopter. I felt sick and dizzy. I was desperate to find out whether the pilots were OK.

For several hours I roamed this ghostly, godforsaken island. I hadn't even meant to come to North Wales, I thought with a sigh. I was sitting down on a rock to work out my next move when suddenly my radio crackled again and I heard the welcome words, 'Stand by for rescue by the RAF.' Moments later a light appeared in the sky, followed by what appeared to be a vast yellow canary, and I heard the distant thud of rotor blades. I was saved. It was a Sea King, a great behemoth of a helicopter that looked more like a flying double-decker bus as it flew past and the pilot assessed the island for a suitable landing spot. Being much larger than the Bolko with an immense downdraught, it required a much larger area. I watched as the pilot battled desperately to maintain its position some fifty feet above me. The noise of the rotor blades was almost as startling as the pilot's ability to keep the machine steady in the storm. Suddenly another figure appeared at the helicopter's open door and he was winched down, spinning in the gusting wind before landing with a thud on the slippery rock.

'Are you all right?' he asked politely. 'We're going to have to winch you. It's too dangerous to land.' The wind had increased and the visibility had been reduced still further. The Sea King was flying close to its limit. If the weather continued to deteriorate it would have to return to RAF Valley and come back for us once the weather had improved. A strop was lowered from the helicopter and the winch man placed it over my neck and under my arms, and before I knew what was happening, I was dangling in mid-air, the ground disappearing below me as I was yanked up higher and higher. Two arms were wrapped around my waist and the next I knew I was sitting on the metal floor of the RAF Sea King. The noise was deafening as the winch man handed me a helmet with ear defenders. I was shepherded to a seat behind the pilot, where another member of crew gestured for me to plug a cable into my helmet.

'Welcome aboard!' came a disorientating voice into my ears. I couldn't tell who was talking.

'What happened? Are the other pilots OK?' I blurted into the microphone protruding from my chin strap.

'The helicopter's gone, but the pilots are fine,' came the reply.

It transpired that when the chopper had returned to the *Patricia* after dropping me off, it had been secured by the crew, but as the pilots walked away a vast wave had broken over the stern, sweeping the chopper right over the side and taking the rail with it. Somehow, miraculously, it had missed the deck below. The emergency floats on her skids had failed to keep her afloat and she had disappeared below the waves. A million pounds' worth of equipment lost to the sea. From the open door I could see the *Patricia* below as we circled her several times. There was no sign of the helicopter, just a dirty black slick of oil in the water. I shuddered at the thought. What if I had still been aboard?

I watched, mesmerized and bewildered, as the ocean raced past the open door. Soon blue turned to green as we sped over cattle-filled fields before we began to hover over a school sports field. The pilot gently lowered the Sea King towards the soggy rugby pitch. 'I'm afraid we've got another emergency. We'll have to leave you here,' the pilot said. I jumped out of the door and the helicopter rose noisily and majestically into the grey sky. I was standing ankle-deep in mud in the middle of a Welsh rugby pitch, miles from anywhere, a spaceman in a bright orange jump suit surrounded by a dozen bewildered schoolchildren. I smiled and waved at them as if this was just the normal course of events and then wandered off down a narrow country lane next to the field. I had no money or telephone, so I had no choice but to find the main road and hitch a ride back to Holyhead. A dozen cars swept past, their drivers staring at me as if I was some kind of alien, before one kind soul, or a curious one at least, pulled over and I jumped in.

It had been an extraordinary day and quite an adventure, I reflected, after a pint of ale and a portion of laver bread in the safety of a Holyhead pub (can those three words be used together?). My jaunt on the Skerries was certainly something to tell the grand-children about by the fireside one day. I even made the front pages of the Welsh newspapers the next day.

I have left many islands, around our shores and in the oceans beyond, dreaming of one day returning as a visitor or even as an inhabitant. But I can safely announce, here and now, that the Skerries are one place to which I shall not be returning to pay a visit, let alone to build my island dream home.

# Isle of Man

It was all a little confusing: 'WELCOME TO THE ISLE OF WOMAN' announced a huge billboard on the top of the airport terminal building as the Womanx Air plane descended through the thick, grey cloud that sat over the Irish Sea. During the flight I had read from cover to cover the *Isle of Woman Independent*, in which there was a slightly disconcerting picture of the local rugby team, Deirdre RFC, dressed in pink rugby shirts. The island's Michael Football Club had been renamed Michelle FC. Even the famous three-legged symbol on the island's flag had been doctored to show three pink, high-heeled female legs. Had I missed something? In Guernsey I had been confronted by teams of seven-foot-tall basketball players from obscure islands around the globe; I'd gone to the Skerries and my helicopter had ditched into the sea; now I'd arrived in the Isle of Man and the entire island had had a gender change – and what's more no one seemed to find it strange.

It looked as if I was in for a rather strange few days in this quasi-independent, Crown possession halfway between Cumbria and Northern Ireland, famed for being the home of Norman Wisdom and Nigel Mansell, the birthplace of the Bee Gees and, of course, home to thousands of tax exiles who live their lives as if they were still in the 1950s. For a weekend the island had renamed itself in honour of the fairer sex in an extravagant publicity stunt dreamt up by Yorkie, to publicize its new pink-wrapped, woman-friendly version of its top-selling chocolate bar. Pubs had become 'women only' and a pink bill had been passed by Tynwald, the world's oldest parliament in continuous existence, commanding men to wait on women hand and foot, to lower the loo seat and not to complain when the phone was in constant use.

The island is served by a regular ferry service from both England and Ireland. I have always enjoyed ferries; in fact I often enjoy the boat trip to an island almost as much as the place itself. The 'getting there', that sense of mounting curiosity and anticipation, is a central part of the whole island experience. I have vivid memories of childhood trips to France and the Isle of Wight, but they are recollections only of the ferry journeys: the holidays themselves have blurred into a hazy concoction of baguettes and sticky rock. I cannot recall a single detail about where we stayed or what we did, but my memories of the crossings remain incredibly vivid. Boat travel gives journeys an extra dimension, adding both adventure and a tangibility lost in air or road travel.

I had therefore intended to arrive by sea, but sadly on this occasion a developing winter storm had stranded the ferries and forced me to use Womanx Air. The flight, an hour long, was from the most depressing airport on the planet, Luton. The only positive thing about flying from Luton airport is that wherever you are flying to is guaranteed to feel like paradise in comparison. I can highly recommend it for trips to Chernobyl, Baghdad and Chechnya.

To be honest I had no idea what to expect on the Isle of Man. I had heard and read much about it, but I had never had the excuse or the opportunity to visit. I had been enthralled by Matthew Kneale's novel *English Passengers*, about a Manx crew who headed to Tasmania in the mid-nineteenth century, and I was fascinated by the glimpses it gave into the island's unique culture and history. I knew also that, because of its favourable tax climate, the Isle of Man had become a centre for the international film industry, and I had heard how many locals, and people passing through, had secured themselves parts as extras in a number of well-known films. Perhaps, I wondered, I could land a weekend job as a passing postman or a tree. My mother, Julia Foster, is an actress and I have always fancied the idea of becoming a footnote in cinematic history. Both my sisters have appeared on the small screen in advertisements for Chocolate Buttons and Fairy Liquid, but I have failed to make an acting debut. The Isle of Man, I concluded, would be my lucky break on to the silver screen and a one-way ticket to Hollywood.

Mainly, though, my motivation for coming to the Isle of Man was simply that, like the Atlantic or Everest, it was there. It was nothing but pure curiosity that drove me there, and there is certainly plenty to be curious about on Man, or Mann, or Mannin, or Ellan Vannin, or Mona, as it has been known at various times by various people over the centuries. Today the island is independent from the United Kingdom in all areas except defence and foreign relations. It has its own currency, the Manx pound, and produces its own stamps through the Isle of Man Post Philatelic Bureau; though English is the preferred language, Manx Gaelic is still very much in use, and there are several dedicated Gaelic-speaking schools.

'Hello,' said Daphne from the island's tourism de-partment, smiling as she munched one of the 20,000 pink chocolate bars Yorkie had distributed to the island's womenfolk. Daphne had been one of the masterminds behind the island's re-branding, and she had offered to meet me at the airport and drive me to Douglas, a twenty-minute journey away through the Manx countryside.

'Manannan's wearing his cloak today,' she whispered cryptically as we drove through the drizzle and low cloud over a patchwork of green fields and open country. I wondered whether Daphne was in fact a secret agent and perhaps this was a secret code to which I needed to reply 'Orangutans like poppies' or some such absurdity. Seeing my bemusement, Daphne quickly enlightened me. 'It's the mist that often shrouds the island,' she explained. Manannan, I dis-covered, is a pre-Christian Irish sea god from whom the name of the island is said to derive. Among the many powers attributed to him, Manannan can make one man appear to be a hundred and protects the island from invaders with his cloak of mist. Invaders like me, I supposed. 'Manannan lifted for the Queen's visit,' Daphne announced proudly. 'You obviously don't have any blue blood.' A few moments later she turned to me and implored, 'Say laa mie to Themselves,' as we drove over a tiny bridge. 'You don't want to upset Mooiney Veggey,' she warned. I looked out of the window, a little nonplussed. They certainly had a strange way of talking on the Isle of Man.

Manx folklore is steeped in tales of fairies, gods and goblins. It is said that the island came into being when the Irish giant Finn MacCool ripped up a huge clod of earth in a fit of rage and flung it and a pebble at a fleeing rival. The clod fell into the Irish Sea and became the Isle of Man, and was then shrouded in mist by a sea god. The hole the clod left filled with water to form Lough Neagh in Northern Ireland. To be sure of a pleasant visit to the island, you are advised to say hello (laa mie) to the Little People (Mooiney Veggey) when you cross the Fairy Bridge we were approaching. According to lore, the Little People can take offence easily and should never be ignored or addressed directly. They should always be addressed as Themselves or Mooiney Veggey.

Islanders used to leave abandoned houses untouched as residencies for the fairies, and obliging neighbours not wanting Themselves to go hungry would leave food and drink on the doorstep for them.

On 30 April each year the islanders fix a wooden cross bound with sheep's wool to the inside of their front doors to ward off malicious fairies and that night people blow horns on Peel Hill to banish evil spirits. Superstition still runs deep on the Isle of Man, I thought, as I watched a burly trucker pass in the opposite direction over the bridge mouthing hello to the fairies.

A few minutes later we had left the fairies, the sea gods and the sodden countryside behind us as we entered Douglas. Located in the gentle curve of a bay, Douglas is a grand Victorian resort with a sweeping promenade and a large, historical harbour. Once the focus for trade in salt, herring, hides, soap and beer, the town today is the centre of the island's tourism industry. The Isle of Man was by far the largest of all the islands I would be visiting on my journey, both in terms of geographical size (221 square miles) and population (roughly 76,000 at the last head count), and it struck me once again as we headed into town that on almost every island you visit people have ignored the wide open expanses around them and congregated in towns and villages, preferring to live cheek by jowl rather than enjoy the greater amount of space and freedom available to them a mile or so into the countryside.

The promenade was swathed in multi-coloured light bulbs dangling from an electric necklace like bunting. It reminded me of Brighton seafront with its impressive, though slightly crumbling, Victorian buildings, many of which are currently undergoing redevelopment. Because it is a tax haven, like Guernsey, many of the larger buildings are home to international banks, squeezed between the garish games arcades and old-fashioned tea rooms. The town has an old-world feel about it and indeed many of its visitors are attracted to the island for its timeless charm and slower, anti-quated way of life. Technically, at least, there are still a number of very eccentric laws on the Manx statute books, not least the one that allows a true Manxman to shoot a Scotsman wearing a kilt if he's on the beach. (Note to all visiting Scotsmen: avoid sunbathing in your clan colours – although it has to be said that with a mean August temperature of just 58°F, the opportunities to top up your tan are few and far between on this island.)

'Here we are: Birchfield Villas!' announced Daphne as we swept up the horseshoe-shaped drive of an extremely well-manicured garden towards a pristinely kept building. 'Johnny Depp and Vanessa Paradis stayed here,' she added impressively. Birchfield Villas was, I'm told, the island's newest 'boutique' hotel and its only Five-Diamond-rated residence. A short walk from the promenade, Birchfield Villas was run by Mr and Mrs Plum, who had recently divorced but continued to live and run the hotel together. I quickly discovered that this arrangement led to an unusual domestic and work situation as the couple rarely seemed to communicate with one another.

'What would you like for breakfast tomorrow

morning?' asked Mr Plum in his American accent, leaning through my bedroom door.

Moments after I had given him my order and he had disappeared, there was another knock at the door and Mrs Plum appeared and asked, in her strong Manx accent, 'What can I get you for breakfast tomorrow?'

She disappeared, and moments later Mr Plum was back, enquiring, 'Would you like tea or coffee with your breakfast?'

As if we were in a good West End bedroom farce, Mrs Plum then returned, bang on cue and – I could almost have asked the question for her – 'Did you want coffee or tea, Mr Fogle?' It was like a real-life game of Cluedo, the winning answer being Mr and Mrs Plum in the bedroom with a metal teapot.

The earliest inhabitants of the island are believed to have been the original Celtic people of the British Isles and the common surnames such as Cortlett, Kelly and Quinlain are all versions of those of the original settlers. Viking invasions began in about AD 800 and the isle was a dependency of Norway until 1266. King Orry, who ruled the island in the late eleventh century, established the three important bulwarks of Manx society: the state, a legislative body and a standing army. The Scandinavian system of government has remained practically unchanged ever since.

In 1266 the King of Norway sold his suzerainty over Man to Scotland, and the island came under the control of England in 1341. In 1406 the English Crown granted the island to Sir John Stanley, and his family ruled it almost uninterrupted until 1736. (The Stanleys refused to be called kings like their predecessors, adopting in place of King the title Lord of Mann, which still holds.) The Lordship of Mann passed to the Dukes of Atholl in 1736, but when the island became a major centre for contraband traffic, depriving the British of lucrative customs revenues, Parliament purchased the island's sovereignty. At a stroke much of the population's livelihood was killed off, the standard of living on the island plummeted and a culture of smuggling on the island was spawned that would continue for many decades. No wonder these Gaelic people have not felt any great allegiance to their British overlords.

The contraband business, or the running trade as it was known, was the means by which merchants avoided high British tariffs on imported goods such as tea, tobacco, wine and brandy by importing them legally into the Isle of Man, paying lower taxes to the Lord of Mann, and then 'running' the goods to colleagues waiting along the shore on the British mainland. It is no exaggeration to say that the town of Douglas was built on the wealth generated by the running trade. After the island became British, the Royal Navy and British customs officials patrolled the ports and waters around the island to enforce the new laws. One of the commanders of the revenue cutters was none other than a certain Lieutenant William Bligh, who a few years later found himself cast adrift from his ship, *The Bounty*, by two Manxmen, Peter Heywood and Fletcher Christian.

As with many seafaring communities, the Manx are inherently superstitious. At sea, Manx sailors were not allowed to whistle as this 'aggravated the wind'. Fishermen were particularly superstitious about the third boat leaving port, as it was supposed to be unlucky. When the contraband business dried up overnight, the Manx were forced to find other ways of earning a living and the island soon began to prosper from the herring trade, for which it is still famous today, Manx kippers being sold around the world.

By turning to the silver screen in recent years in order to find a fresh source of income and employment for its people, the island has earned itself the moniker of 'The Hollywood of the Irish Sea'. *The Brylcreem Boys*, *I Capture A Castle* and *Waking Ned*, which were all filmed on location, have all made a significant contribution to the island's economy, while cleverly boosting its profile as a tourist destination by advertising its beautiful landscape for free in the films.

In 2005 alone more than a dozen feature films were shot on the island. David Jason came to Douglas to film *Quest III*, Sean Bean flew in for *The Dark*, Stephen Fry had the huge and highly impressive King William College situated next to the island's runway taken over for the screen adaptation of *Tom Brown's Schooldays*, while Castletown became the location for the television drama *Islands at War*. Not long before I arrived, and much to every 'womanx's' delight, Johnny Depp and John Malkovich spent several weeks on the island making *The Libertine*. Even Bollywood has discovered the charms of the island and its equally enticing tax incentives. This burgeoning demand has led not only to the construction of a huge island studio to rival Pinewood but also to a dedicated casting agency, Nina's People, which has over a thousand extras on its books. As I sat in my room in Birchfield Villas, watching the weather worsen by the minute and feeling the central heating rising with every degree as the temperature fell outside, I flicked through the phone book and started dialling.

'Hello, Nina's People,' announced a voice cheerily.

'Er, yes, good morning, I'd like to be in a film,' I stammered.

'Doesn't everyone, darling,' came the curt reply. 'Just fill in an application form from our website and send it back to us with a £25 registration fee and we will keep you on our records.' Her tone sounded mildly promising, and I knew that Bob Hoskins and Brenda Blethyn were meant to be on the island filming *On a Clear Day*. I sniffed a chance and raced to the nearest Internet café to log on to the website.

'So you want to be in the movies?' asked the site. 'Have you ever fancied being in films or on TV? Well, now you can, by becoming an extra, one of the lucky people who get to mix with the stars . . .' Yes, yes, yes, I muttered at the screen as I conjured up images of me, a blond James Bond, sprinting across the backs of alligators, riding helicopter skids and bedding Ursula Andres. 'To become an extra all you need to do is send in the application form together with two photographs and £25,' the site explained. The daily payment rate was £65 for a twelve-hour day or £82 for a night shoot, with 10 per cent deducted by Nina for her troubles – not bad work if you could get it. I later discovered that Mr and Mrs Plum, Daphne and even the chippie from whom I had bought my fish and chips were all registered extras.

I hurriedly noted down my eye colour, height, hair colour and inside leg, and then I reached the section entitled 'Previous film experience'. I had precisely none, but my mother had appeared in several films including *The Great McGonagall*, with Peter Sellers and Spike Milligan, a film inspired by William McGonagall, a nineteenth-century Scot widely regarded as the worst poet of all time. (Who can forget the immortal lines from his woefully bad 'Tay Bridge Disaster': 'Beautiful Railway Bridge of the silv'ry Tay!/Alas! I am very sorry to say/That ninety lives have been taken away/On the last Sabbath day of 1879/Which will be remember'd for a very long time.') While my mother was filming I had been a developing foetus inside her bulging tummy. That seemed good enough 'previous film experience' for me and so I eagerly tapped in *The Great McGonagall* with a flourish and 'with Peter Sellers' for added impact. Then came the killer blow to the launch of my Hollywood career: 'Please allow several weeks for your application to be processed,' the site informed me after I had completed the application.

Plan A for the Isle of Man was down the drain and so I stepped back into the wintry squalls outside and skulked down the high street to the promenade, where horse-driven trams still ferry tourists up and down the seafront. It was like stepping back in time, or indeed into a film set. History moves slowly on the Isle of Man. Although the island gave the vote to women as long ago as 1881, homosexuality was only made legal in 1992, and the death penalty remained on the statute books for another year. There are no licensing laws on the island and, somewhat alarmingly, there is no speed limit – hence the world's most famous motorcycling race, the annual TT race.

The origins of the TT can be traced back to the last years of the nineteenth century when the American James Gordon Bennett, son of the owner of the *New York Herald*, established the world's first motor race with the Gordon Bennett Cup Race. Britain at the time had a strict speed limit of fourteen miles per hour and so the Isle of Man with its independent Parliament and fledgling tourist industry stepped in with an offer to stage it. Over a hundred years later, the meeting has developed into the two-week long summer TT festival, attracting thousands of spectators and riders from across the world. Stanley Woods, Geoff Duke and Joey Dunlop have all tasted TT success and ultimately sacrificed their lives on the thirty-seven-and-three-quarter-mile course that requires absolute concentration and an almost photographic memory of the course. In the 1980s, another famous motorist, Nigel Mansell, moved to the island for tax purposes and became one of the island's volunteer traffic policemen.

Local lore still recounts the times islanders were chased by the Formula One driver and prosecuted by PC Mansell.

Compared to the other islands I was visiting on my offshore adventure, the Isle of Man was positively enormous, at thirty-one miles by eighteen virtually a continent. It would take me a week at least to explore it by foot or bicycle and so I rented a car and set off from Douglas into Man's hinterland. My first stop was Niarbyl, a little picture-postcard settlement with whitewashed thatched homes and a red telephone box, now famous as the setting for *Waking Ned*. A tiny path leads down to an idyllic stone cottage on the water's edge with lobster pots stacked neatly along its wall. A small brightly coloured fishing boat bobbed on the waves as the Irish Sea swelled around it. No one lives in this beautiful spot. It's all seemingly for cinematic show. Looking around, it was easy to understand why the Isle of Man has become such a popular film location. The countryside and coastline are unspoiled by modern life and thus perfect for historical settings, but they could also double up as any number of places around Britain and Ireland. One minute you can be in a tiny remote Irish village, the next in the highlands of Scotland or the Yorkshire Dales; and if you head down the road to Castletown, with some clever camerawork and a bit of imagination you could be in eighteenth-century London.

My next stop was Dalby, the reputed home of a talking mongoose. On an island with a greater cast of fairies, ghouls, goblins and monsters than *Rent-a-Ghost* (does anyone remember the television show?), it is the story of Gef that seems the most incredible and yet credible. The story dates back to the 1930s, when a family called the Irvings lived in a remote farmhouse in Doarlish Cashen near Dalby. The family had travelled the world for many years, before eventually settling on the island. James and Margaret were struggling financially to provide for their three children and James had taken it upon himself to repair the draughty, damp house to protect it from the bitter elements. One day, while insulating the cold stone walls, James noticed a noise coming from behind the wooden panelling. The noise became more and more bizarre and when James started imitating a cat, thinking it might be a mouse, the noise began to imitate him imitating a cat. The noise turned out to be not a cat, nor a mouse, but a talking mongoose named Gef, with whom, it transpired, their daughter Voirrey had been having animated conversations in French, German and even Spanish. As if a multilingual talking mongoose wasn't odd enough, Gef had explained to Voirrey that he was not in fact Manx but Indian – he had been born in Delhi in 1852.

Gef, who would always appear by singing the Manx national anthem, soon became a part of the Irving family, tormenting and thrilling them with his gossipy tales about their neighbours, on whom he would spy, and even catching rabbits for the Sunday lunch pot. He never let them touch him and once he even bit

James for trying to stroke him, afterwards announcing grandly that 'he should put some ointment on that'. Unsurprisingly, the story of Gef the talking mongoose soon reached Douglas and even mainland Britain. A Hollywood agent is reputed to have offered the family £30,000 in cash for the film rights, and reporters flocked to Doarlish Cashen.

The story caught the public's imagination and even attracted the attention of Harry Price, a famous ghost hunter, who spent time with the Irvings but failed to meet Gef, though he claimed to have some grainy photographs which were never authenticated. The remote farmhouse soon became the focal point for curious sightseers and ghost hunters on 'pilgrimages' to meet the talking mongoose, who once announced grandly that he was the 'fifth dimension, the eighth wonder of the world'. Despite all this attention and their new-found fame, it seems the family never capitalized on being under the spotlight. Indeed they were eventually driven from their home by the army of visitors, who labelled them mentally disturbed, and the story of Gef was forgotten – until, that is, several years later, when a local farmer shot and killed a strange-looking unidentifiable furry animal on his property.

The family disappeared, the story became folkloric and Voirrey refused ever to talk about Gef again, claiming that he had indeed existed but that he had ruined her life and that she wished he hadn't. What made the story all the more tantalizing was the discovery some years later that a neighbouring farmer had once imported some mongoose from Delhi to keep down the island's rabbit population, and that Indian lore speaks of talking mongooses. The tale still attracts attention, although the Irvings' farmhouse is long gone.

Perhaps unsurprisingly there was no sign of Gef as I wandered through Dalby, but, if nothing else, the story illustrates the extraordinary gift the Isle of Man has for producing tales and fanciful fables. Like many islanders, the Manx are full of tales, but their imagination seems to be far more fertile and lively than most. Perhaps it is a case of the bigger the island, the greater the scope for the imagination to roam.

As I wandered through the chocolate-box village a cat hopped on to a wall next to me and began grooming itself. I noticed that it was missing its tail – there was not even a stump. As a child I had watched my father, a veterinarian, perform dozens of amputations on cats' tails that had been caught in car doors and gates, but there had always been a stump left. A young woman came along and scooped up the cat.

'What happened to her?' I asked. 'It must have been awful.'

'What do you mean awful?' she answered with shock. 'She's an award-winning pedigree cat. She's not meant to have a tail,' she added, before turning on her heels in disgust.

In fact I was looking not at an unfortunate amputee but at one of the island's most famous inhabitants, the Manx tailless cat, the first sightings of which were recorded on the Isle of Man in the early nineteenth

century. One theory, typically Manx in its outland-
ishness, is that two cats were the last to board Noah's
ark and had their tails severed by Captain Noah when
he slammed the door on them in his haste to beat
the weather. Another hypothesis is that they swam
ashore from a sinking vessel of the Spanish Armada in
1588. Some islanders blame the Irish, whom they
accuse of cutting off Manx cats' tails and using them as
helmet decorations, while, more prosaically, scientists
believe the taillessness is nothing more than a mutant
gene.

'There's a severe weather warning in place on the
island,' warned Manx FM as I manoeuvred my little
Ford Focus round the island's narrow lanes. 'Islanders
are advised not to go outdoors unless absolutely
necessary.' The wind had picked up considerably since
I had left Douglas but I didn't have time to stay
indoors: it's essential research, I told myself as I drove
on heedlessly towards the Calf of Man, a tiny island in
the south-west separated by the Calf Sound. The island
is now a bird sanctuary and nature reserve, and also
the location of Chicken Rock, famous for its lighthouse, which was built in 1885 in an effort to reduce
the number of ships wrecked on her treacherous
shores. The sea had been whipped into a wintry frenzy,
sending waves crashing across the Calf. It was difficult
to stand in the powerful wind blowing in off the Atlantic and up the Irish Sea, which had begun wreaking
havoc across the island. This was no ordinary storm.
'Gusts of 100 miles per hour have been reported in

the south of the isle,' reported the announcer breath-lessly as I felt my car being buffeted by the wind. The huge trees that lined the road swayed dangerously, as branches and leaves rained on the deserted highway.

Foolishly I drove on, and as the falling debris increased in frequency and size I was forced to drive around large branches that were partially blocking the road. I watched helpless as a tree in front of me collapsed, creaking as it toppled with an almighty thud on to the road just a few dozen yards ahead of me and creating an impassable obstacle with its enormous trunk.

'Blimey,' I mumbled nervously. Suddenly I didn't feel so brazen, and I turned the car round and sought another route back to Douglas. My heart was pound-ing as I made my way through the worsening storm. The road had turned into a river of branches and leaves and I manoeuvred my little hire car around the debris, often on to the grass verges to avoid the larger branches that had crashed to the ground. 'The Civil Defence has now been called out. Roofs have been blown off and hundreds of trees have come down,' the radio presenter informed me with an even greater sense of urgency in his voice.

For six hours I navigated my way across the isle like Pacman, my passage frequently blocked by an ever-increasing number of fallen trees. Locals with chainsaws braved the elements, trying to keep the island's infrastructure open. Power cables dangled precariously across the road, felled by falling trees, and the island's railway line had been brought to a halt by hundreds of uprooted trees lying across its tracks. 'Meteorologists confirm that this is the worst storm to hit the island in twenty years,' the radio presenter announced as I finally limped back into powerless Douglas, its houses illuminated by a few paraffin lamps and candles. I collapsed on to my bed exhausted, and fell asleep to the lullaby of the wind.

'That was quite a storm. I hope you slept,' said Mr Plum before disappearing into the cellar.

'That was quite a storm. Did you sleep?' asked Mrs Plum as I tucked into my eggs and bacon.

It was Groundhog Day, and my final day on the Isle of Man, which had been all but closed down by the ferocious weather. Half the island was still without electricity and television reception. Greenhouses had been smashed, pigsties blown away; dozens of homes no longer had roofs, and more than five hundred trees had fallen.

The storm had subsided but the sea was still bubbling with anger, and the sky was streaked with an ugly rainbow of greys that merged into the Irish Sea. A dank blanket of drizzle soaked my jacket as I sloshed through puddles into a brisk headwind on a final walk along the promenade. A man walked his dog and several 'hoody'-clad teenagers cycled past and splashed me with water.

A couple walked towards me, their arms linked tightly. He wore a thick raincoat and she a long trench coat, and they both wore peaked caps. They were both strangely familiar, I thought, as I struggled to place their faces. Then it dawned on me: I had seen hundreds if not thousands of pictures through the years, and now, on this strange island, in the middle of a storm, here were the real-life Mr and Mrs Ritchie, Guy and Madonna. Of all places.

My head went fuzzy. Madonna was part of my childhood; I'd practically grown up with her. I had read on the Nina's People website that Guy Ritchie was on the island making his new film, *Revolver*. This is my chance to become an extra, I thought, sounding more and more like Ricky Gervais from his BBC2 television series.

The production had nearly finished but surely there was scope for a posh gangster? I was assuming that his latest picture was going to be another *Lock, Stock and Two Smoking Barrels* East End gangster film. Surely I could persuade him to do a little rewriting? After all, he'd given Madonna a scene even after the critical vilification of their disastrous co-production in *Swept Away*, and I had a screen credit of my own, albeit an invisible appearance in *The Great McGonagall*.

What's more, I had even met Madonna once, or should I say, I had seen her once. I was working for a silver-service party organizer in London while studying at university. The job was poorly paid but it entailed serving champagne and canapés at all the best parties and film premières, including *Evita*, in which Madonna was starring. I had been so starstruck that I'd attempted to fill her glass with an unopened bottle of champagne, much to her amusement. My friends and I had spent the entire evening trying to get her attention, and then at three in the morning, when our shift was over, we had all dashed out on to the red carpet as she left, hoping to get snapped by the photographers, only to be held back by several burly bouncers, dashing my chance of becoming Mr Madonna.

I wondered whether she'd remember me. The couple were just a few paces ahead of me. How should I address them? Mr and Mrs Ritchie? Mr and Mrs Madonna? Guy and Madonna? Madge? 'Can I be in your film?' was all I had to ask, I thought, as my mouth went dry. Just six words.

'Hello,' I said, grinning foolishly, my heart pounding. What had come over me? I was acting like a schoolboy.

'Hello,' she replied, her smile lighting up the dull sky. I swore I heard the fairies sing. And that was it: with a single hello they disappeared into the Manx mist, leaving me speechless and filmless. Once again I'd failed to make an impression on the Material Girl.

It was time to move on. My rainswept stopover on the Isle of Man had ended as strangely as it had begun, but in a superstitious land of talking mongooses, tailless cats, demanding fairies, pink chocolate bars, ancient sea gods and bizarre hotelkeepers, 'strange' seemed entirely fitting. My weird weekend on the island had been every bit as odd as the fabulous folklore that is woven into the culture. I had enjoyed it, but for me the Isle of Man was just a little too strange to be my dream island. The search would have to continue.

# Muck, Eigg and Gruinard Island

## *Muck*

Muck may have a slightly unappealing name but the island, twelve miles from the Scottish mainland at the southern end of the Highlands and the smallest of the Small Isles, couldn't have a less appropriate one. The unusual name derives from the Gaelic word *muic*, meaning island of the cow. An earlier owner, who disliked the name, is said to have tried to convince Samuel Johnson and James Boswell, on their celebrated tour of Scotland, that its real name was the Isle of Monk.

This small patch of pastureland in the Sea of the Hebrides with its thirty inhabitants is a working farm in all but name, run by a laird who looks more like someone from the Battle of Bannockburn than monarch of the glen. I had seen him once before in a television car advertisement playing an RNLI dentist rushing out on call, and I had been struck by this vast Scotsman with his great red beard, barrel-like chest and arms like giant hams. 'That man owns Muck,' a friend had whispered. My journey around our coast presented me with the perfect opportunity to meet this extraordinary character, and visiting Muck fitted in nicely with my original plan to stay on neighbouring Eigg, which had endured some bizarre characters of its own since my last visit there as a young boy.

Getting to the Small Isles is not easy. A flight to Glasgow was followed by a six-hour drive to the little Highland harbour town of Glenborrodale on the southern side of the Ardnamurchan peninsula, a tiny two-house town hidden down a winding single-track road. After a night in a B&B I boarded the small boat that ferries the inhabitants of the Small Isles to and from the mainland. Each week it collects older schoolchildren from various islands, and they board for the week in Tobermory on the Isle of Mull and attend the High School. Tobermory has recently found fame as the location for the fictional town of Balamory, in the children's television programme of the same name. Interestingly the town is now a booming tourist destination, but not for the ramblers and hikers endemic to many Scottish towns but for hordes of six- and seven-year-olds who descend on the town in thousands. So great is the 'children's invasion' that hotels have been forced to buy extra children's beds and restaurants have had to alter their opening times and even menus to cater for the young *Balamory* groupies.

The boat journey from Glenborrodale takes roughly eight hours, depending on the weather. You pass through some of the most spectacular coastal scenery around our shores as you head out into the open sea with Mull to the south before you turn due north up the coast to the Inner Hebrides. Skye's snow-capped Cuillin hills soar above the green sea.

Muck differs from the other Small Isles – Rhum, Eigg and Canna – in that most of it is very low lying. The first thing that struck me as we approached was the scattering of wind turbines along the island's coast-line. I have always been a champion of wind turbines. They are beautiful, both aesthetically and environmentally. I have spent time living next to wind turbines and truly believe that they enhance rather than detract from the scenery. I can still remember the first time I saw a wind farm, and it truly took my breath away (pardon the pun). I find the whole 'not in my back yard' (NIMBY) mentality wildly hypocritical and short-sighted. Muck's wind turbines were beautiful too, and I delighted in the organic lullaby of their swishing blades, beating in time to the ocean. I used to fall asleep to this fantastic eco-friendly music when I lived next to wind turbines.

The people of Muck, I would soon discover, are a very resourceful, self-sufficient and practical bunch who do what is reasonable to sustain their enviable way of life. As our boat approached the little harbour of Port Mor we were met by the rather surreal sight of dozens of Highland cattle and other cattle roaming the white sandy beach next to the harbour and munching from hay bales while half-a-dozen children played among them and built sandcastles.

The island is owned by two brothers: the Laird of Muck, Lawrence MacEwan, and his younger brother,

Ewan MacEwan, to whom it had passed down the family after it was bought in 1896 by their great-uncle. Robert MacEwan, or Robert Thomas, as he was also known, was a *Times* correspondent and arms dealer who supplied the military hardware for revolutions in Peru and Chile as well as uprisings in Afghanistan and China. He bought the island on the proceeds of sales of battleships to Japan. It cost him £15,000.

Today the island is home to thirty residents who farm the land and waters that belong to Laird Lawrence. Like many other isles around the UK, the tiny island, just two miles by one mile, had been struggling from debilitating depopulation as islanders sought an easier life on the mainland – until, that is, Lawrence and the island committee decided, like Bardsey, to put an advertisement into the papers calling for new blood. The local primary school faced closure as its numbers had fallen to just two pupils. The appeal for a family to move to the island resulted in more than two hundred applications, from as far afield as New Zealand, most of them young families willing to abandon their city life for the good life.

The committee, headed by Lawrence, whittled down the applicants to six families, who were each invited to stay on Muck for between three and seven days to meet the residents. The adults on Muck then voted for whom they would most like to have living with them.

The lucky family chosen by this very transparent and democratic process were Nick and Jill Noble and their daughters Caitlin and Ellish from Leeds. Inspired by the BBC's television drama *Two Thousand Acres of Sky*, set on a fictional Scottish island, and by my own adventure in BBC1's *Castaway*, the Nobles agreed to swap their urban life for life on an island with no shops or pubs and accessible from the mainland just three times a week – weather permitting, of course. They were allocated a three-bedroom house at a rent of £240 a month plus £1 a month for electricity from the turbines. As a freelance computer analyst, Nick planned to work from home and his wife, a former health authority worker, was offered work in the island's school.

Nick was waiting for me at the jetty with his young daughters as I stepped off the little boat, a bit wobbly on my feet after eight hours of bobbing up and down over the waves. 'Welcome to my home!' he said proudly, beaming and opening his arms in a theatrical gesture.

I had met Nick a number of times before and seen him many times more in the newspapers and on the television. The story of a young family moving from Leeds to a remote Scottish island had captured the public imagination, and there had been dozens of newspaper articles and television appearances detailing their experiences. I had interviewed him and met him on the set of a number of breakfast shows and daytime programmes. 'I'm afraid I'm off on the next ferry. I've got to go to Manchester for a chat show. I'm on telly more than you now,' he said, giving me a wink and a smile. His frequent absences, I would soon find out, hadn't gone unnoticed by the islanders, who had begun to question his appetite for life in the limelight.

'Come up to the house for a cup of tea before I leave,' he said before going up the hill towards the little cluster of houses that form the island's community. 'It's the fourth house on the right,' he hollered as he vanished over the brow. On an island with only a dozen homes, there is no call for addresses. A few Land-Rovers and tractors provide transport around the tiny island, which is given over entirely to farmland for the cattle.

A huge bear of a man approached in grubby red farmer's overalls. He had wild ginger hair that seemed to merge into an equally untamed orange beard. 'Hello,' he said shyly, his eyes looking away, offering me the largest hand I had ever seen in my life, a spade of a paw twice the size of mine, worn and calloused from years of hard work. 'I'm Lawrence,' he said as my hand disappeared somewhere into his and he squeezed it with an iron grip. He was certainly a very striking figure to behold, a modern-day clansman in every respect, lacking only a claymore and kilt to complete the caricature.

Lawrence told me that he had always assumed that his brother would run the island, but Ewan had been terrified of not finding a wife, so after studying agriculture on the mainland, Lawrence had returned to the island. 'I still managed to find a wife, though, so I got the best of both worlds,' he said. 'She lived on Soay, twenty miles over the sea from here.' The couple had courted by commuting across the treacherous divide.

Lawrence may be a laird, one of the last feudal overlords in existence, but there were no airs and graces about him as he stood there in his red overalls, which were covered in the evidence of a hard day's work, smeared in oil, dirt and manure. He might have looked like a man eager to bury an axe in your head, but he radiated a great personal charm and a sense of humility. He was certainly not a man to stand on ceremony or suffer fools gladly. I heard a story from one islander that a yacht had once anchored a short distance off the island. A man and woman had come ashore and politely asked Lawrence whether they could possibly use a shower on the island as their yacht had suffered technical difficulties and they had gone for several days without. Irritated by the request, he told them that the island was suffering from a water shortage and they were all far too busy to be able to help. The couple apologized for troubling him and returned to their yacht. It was only once they had departed that one of the islanders pointed out that the couple were none other than Princess Anne and her husband of the time, Captain Mark Phillips. Lawrence simply shrugged and returned to his work. A water shortage was a water shortage. The story is a good illustration of the very obvious affection he holds for the island and the people he lives amongst: their needs were paramount, and not even royals were going to compromise them.

When I asked Lawrence how he thought the Nobles were getting on in their new life, he replied, 'We have had families who have tried to live on the island before.

They find it very hard, especially living in a small community like ours. It can be like living in a ship. You have to get on with everyone and if you don't things can be quite tough.' The main reason that families had decided to leave was the lack of a secondary school, with pupils having to travel to the mainland and stay in lodgings.

Lawrence excused himself, as he had to return to his cattle, and I headed up the hill to the Nobles' new home, a chocolate-box house with panoramic views out to sea. Their young children played on a climbing frame in the little garden outside. It all looked blissfully ideal, I thought, as I knocked on the door. Inside, the walls were plastered with newspaper cuttings about the couple. I asked Nick whether life on Muck had met their expectations.

'When we arrived we were accompanied by five film crews,' he said, his eyes beaming with pride.

'Not the attention, but island life,' I pushed.

'Freedom, space and time,' he answered rapidly, as if on autopilot. 'There are no traffic jams and if I want to visit a friend I just walk,' he continued. 'The island gives me the flexibility to work as a computer analyst online while enjoying the countryside and the wildlife offline.'

'Nick, it's the *Daily Mail* on the phone,' cried Jill from the kitchen.

Nick grinned and disappeared for another interview. I suspected computer analysis wasn't his only line of island work.

I could see why the Nobles had been attracted to the place; I too was taken with the island's simple beauty. The sun beat down on the tiny island dominated by Benn Airein, a 500-foot hill offering spectacular views towards Eigg and Rhum and to Skye's Cuillin hills in the far distance. Islanders pedalled past on bicycles while tractors worked the fields. It was a picture-postcard scene of rural bliss, as grey seals basked in the sunshine and young children ran carefree. There is an abundance of wildlife packed inside Muck's short shoreline and naturalists come here to see rare butterflies, otters and the large colonies of seabirds that nest here. The last reported crime was the theft of two bottles of wine over twenty years ago. The theft was witnessed by one of the neighbours, so the local criminal didn't get away with it – but he escaped with a good old-fashioned telling-off.

Naturally in a community of just thirty, each inhabitant holds more than one job. A man who held more than any was Brian Walters, who was the island's fisherman, boatman, ferryman, coastguard, postman and builder – an impressive portfolio for a man who had moved with his family to the island as a part-time boatman thirty-one years earlier. Brian had invited me to spend the afternoon in his fishing boat and as we set off across the lightly choppy waters, he told me of his great love for the island, and his affection for the big bear of a laird to whom he paid his rent.

'Lawrence isn't your traditional feudal landlord. He lets you get on with your life and rarely interferes,'

Brian explained as we bobbed up and down on the waters while he emptied and dropped lobster pots. His two grown-up children had both flown the nest and the island and he confided that he hoped they would one day return. 'It's not an easy place to live when you are young and looking for love,' he admitted, adding that it was the island's transport links that were the main problem: the island is often cut off for weeks at a time in the winter months.

There is a ferry service from Mallaig on the mainland's west coast, which brings groceries, mail, passengers and vehicles three times a week. A new £6 million ro-ro ferry terminal had recently been opened on Muck and though it appeared to be a great improvement on the previous system, on average it missed one call in three. 'That makes life very difficult for people who travel for business,' he said. The island was home to a number of people with work commitments and interests on the mainland, including an architect – and of course Nick, the island's doyen of newspapers and television studios. 'Nick is enjoying the fame a little too much,' Brian smiled. 'On Muck you're either here or you're not, and Nick spends more time being a celebrity than an islander.'

It was easy to warm to Brian's mild manner. Like Lawrence the Laird, he didn't seem to ask too much from life. He cherished Muck and he loved the life at sea just as much. It had been a memorable afternoon spent in the company of a genuinely nice man.

Not long after my stay on Muck, I was horrified by

the dreadful news that Brian had drowned in the waters he loved so much when his boat sank not far from the spot where we had reflected on his happy existence on the island. Shortly after that, the Nobles left Muck too. For them this island paradise simply wasn't flexible enough for the lives they wanted to lead. It was a double blow for an island with a declining population, facing a constant struggle to sustain a society and maintain its traditional existence. On the face of it, Muck seems to have everything going for it as an ideal island on which to live. It is small yet spacious, remote yet accessible, and very beautiful. It has a tight-knit, happy community with a village hall that doubles as a meeting place or pub when the islanders get together, and it is run by a kindly, open-minded, hard-working laird with their best interests at heart. I had been smitten by Muck, but I was profoundly affected by the news of Brian's untimely death and of the Nobles' departure. Both were stark reminders of the fragility of island life.

## *Eigg*

All was not well on Eigg. Five years earlier the island had become a powerful symbol of independence and self-determination when the inhabitants began to govern it themselves, having freed themselves from the quasi-tyranny of life under a sequence of lairds who might charitably be described as a little eccentric;

the press dubbed the island 'the People's Republic of Eigg'. When I arrived, however, the community was reeling from a string of scathing newspaper attacks. The most recent furore had come about after a German journalist, Reiner Luyken, had written an article for the German broadsheet *Die Zeit* in which he had compared the island to George Orwell's novel *Animal Farm*, implying that the former lairds had been the 'men' and the islander trustees 'the pigs'. It also portrayed some of the islanders as characters out of *The Wicker Man*, suggesting they were drunken, drug-addled and menacing, and had a penchant for suspect sexual behaviour.

The islanders were especially irritated because the author was a well-known enemy of the Eigg Heritage Trust, the body that now owns and runs the island. A well-known opponent of land reform, he lived just down the coast at Achiltibuie and had penned many excoriating denunciations of the new laws that allowed island people greater control over their lives. The article had created a stir on the island but more significantly it had renewed unwanted interest in it. Such interest was fuelled by a former Laird of Eigg once proclaiming, 'Eigg is run by people who lived in Tibet and have "Make Love Not War" painted on the side of their vans.' Life has never been black and white on the most famous of the Small Isles, an island with a colourful cast of characters and one of Scotland's most picturesque settings.

I had spent the night on Muck, sleeping on the little

ferry boat in the harbour, and the next morning we set off northwards on the two-hour journey to Eigg. As we approached, I was struck by how imposing the island was. Although it is only five miles by three, the mountains created the illusion of it being far larger, perhaps because you couldn't see the sea across the other side. It was difficult to imagine that its beautiful white sands were once used by commandos training for the Normandy landings. Known as the island of flowers, Eigg boasts an extraordinary abundance of wildlife, and draws hundreds of naturalists and ordinary visitors to enjoy its unspoilt beauty. Otters hunt and swim around the sandy bays, while corncrakes can be seen standing in meadows rich in orchids, swaying grasses and rare butterflies. Further inland, lochs punctuate the moors and volcanic crags with eagles circling above, while the woodlands are home to a variety of songbirds. Out at sea seal colonies are a common sight, and dolphins and minke whales make frequent visits. You wonder how such a small island can contain such an astonishing diversity of habitats and species.

As our boat rounded the corner to its landing place on the west of the island I felt a great swell of nostalgia rise up in me. I had first visited Eigg when I was eight. My sister and I had been taken there by our Scottish nanny, Katrine, for the Easter holidays. I still have vivid memories from my first proper island experience. I remember leaving Mallaig in a tiny fishing boat and sailing past hundreds of grey seals, the first I had ever seen. It had been a very stormy crossing and we transferred into a tiny open rowing boat to be ferried ashore. We stayed in a tiny cottage in the middle of moorland dominated by enormous cliffs, and my sister and I spent our time variously up to our waists in rainwater, playing cards and listening to the Beatles on my new Sony Walkman. The wind blew relentlessly and the rain lashed us horizontally for the whole two weeks we were there. Time, of course, has eroded many memories, but I recall the Laird's 1920s Rolls-Royce and his eight-wheel caterpillar car that could drive over the roughest of terrains. Our nanny had once looked after the Laird's children, and we spent much time in their grand house. I even spent an afternoon fly fishing with the Laird; I still have a tweed deerstalker hat he gave me as a memento of my trip. I remember watching the island's enormous peak, An Sgurr, disappearing into the low cloud as our boat sailed away and how curious my friends had been on learning I had spent Easter 'on an Egg'. Now twenty-three years later as the island loomed ever closer I wondered whether that childhood trip had contributed to my obsession with islands as well as my Canadian holidays.

I returned to a very different island, physically unchanged, but politically and socially altered beyond recognition. Currently home to seventy, the island had once been a thriving settlement of 500 producing potatoes, oats, black cattle and kelp. The kelp industry, which involved harvesting the seaweed that washes up on the shores, financed the building of the main farmhouses on the island, until the chief's policy of raising rents caused many of the islanders to emigrate to Canada. With the Clearances of the eighteenth and nineteenth centuries, which started because better prices were offered for land empty of people where sheep could be pastured, more islanders were forced to leave, gradually depopulating the island further.

Eventually the island was bought in 1896 by Laird Lawrence's great-uncle, who had also purchased Muck. Subsequently it was sold to a succession of magnates and after falling into the hands of Walter Runciman, a politician and shipping magnate, a character called Keith Schellenberg, a wealthy businessman and sports enthusiast from Aberdeenshire, bought the island for £250,000 in 1974. The former Olympic bobsleigher electrified islanders when he flew in to pay his first visit as their new overlord. He left his plane in the care of a bewildered islander, who was instructed on which buttons not to press. Said buttons had promptly been pushed and the plane was a write-off. The new ownership had started with a bang.

During his early years as owner, Schellenberg banned hunting and declared the island a wildlife preserve, reopened the community hall, converted abandoned crofts into holiday homes, and hired an assortment of artisans and labourers. The population doubled. But Schellenberg also issued strict dictates. When he briefed new staff, he informed them that alcohol and pop music were prohibited from the Lodge where he lived, and because he was a vegetarian, meat was never to be cooked where he could smell it. He unsettled some of the islanders in 1991 when he told the *Highland Free Press* that he had kept the island's style 'slightly run down – the Hebrides feel', yet he also spent time in St Moritz and launched the Eigg Games, a largely social event, pictures of which appeared in glossy society magazines.

By the early 1990s Eigg was more than 'slightly run down' and the costs of Schellenberg's divorce from his second wife left the island in financial limbo. Tenants could not negotiate leases and the estate farm suffered. When Schellenberg finally recovered the island from debt collectors in 1992, its people were naturally suspicious, and some even became openly hostile. When a mysterious fire destroyed his beloved vintage Rolls-Royce, all-out civil war erupted. He called the islanders 'rotten, dangerous, and totally barmy revolutionaries' and they responded by describing him as 'a mean-spirited playboy' and 'a landlord in the worst tradition of nineteenth-century feudalism'.

Soon afterwards Schellenberg sold Eigg and re-turned to collect what he considered his belongings. Among these was a map of the island dating back to the early 1800s that previous lairds had treated as part of the island's patrimony, or so the islanders claimed. To prevent him from removing it, they blocked the door of the storeroom where it was kept with an old bus. Journalists alerted to the confrontation arrived by helicopter. A policeman who was already on Eigg to check its polling station told Schellenberg that he could not move the bus for thirty days. The islanders cheered Schellenberg's final departure, when as his boat pulled away from the dock he shouted at them rather pathetically: 'You never understood me. I always wanted to be one of you.'

The new Laird proved to be even more colourful than his predecessors. No laird can have entertained more fanciful dreams for his land than Marlin Eckhard, or Maruma, as he preferred to be known. (Apparently he had seen the name appear in a puddle during a vision he experienced while in Abu Dhabi.) Fire and wombs fascinated him, and the press reported that he had bought the island because the opening of a local site known as Massacre Cave resembled a giant vagina. Arriving by helicopter, Maruma promised to build a bakery, brewery, distillery, swimming pool and pier-side complex with coffee houses, shops and even a 200-bed hotel. He spoke of more ferries, wind turbines and solar power, and an Isle of Eigg bank that would give low-interest loans to inhabitants. He vowed to build a new community hall, something the islanders wanted very badly, and they hired an architect with their own money to design it. He told a Scottish paper that Eigg would become 'a pilot in self-sufficiency and efficient practice that will be an inspiration'.

On a second and even briefer visit, he ordered everyone to gather the island's abandoned cars on the pier and promised to send a barge to remove them. It never arrived. When he next ordered the islanders to remove all the sea kelp from Eigg's beaches, on the grounds that it was unsightly, the islanders were plunged into despair. Maruma was a 'fire artist' from Stuttgart. His scorched canvases were sold to sheikhs for vast sums of money, but after his shortcomings were exposed the bottom fell out of his 'fine art' business. He already had financial woes and this exacerbated them. Instead of creating jobs, Maruma cut them by selling off the estate cattle. His academic credentials turned out to be suspect and it was revealed that he had borrowed heavily to buy Eigg, and then immediately pledged the island as security for a £300,000 loan from Hong Kong businessmen. Just fifteen months after buying the island for $2.5 million he put it on the market for $3.2 million.

A bizarre twist in the Maruma story came when a German consultant, Dr Kals, approached the Highland Council. Claiming to represent Pavarotti, the great Italian tenor, he said Pavarotti's foundation was interested in building a European centre of excellence for 3,000 classical music students on Eigg. The plans were eventually dropped because of a lack of facilities.

Within a year, Eigg's 'barmy revolutionaries' had bought it from its creditors. A major driving force and inspiration in the struggle to free the island from the whims of passing tyrants was Maggie Fyffe. Eilean nam Ban Mora – the Island of Big Women – had lived up to its name.

I went to visit Maggie in her modest bungalow with views stretching out into the Sea of the Hebrides. Her kitchen was cluttered with piles of correspondence and dozens of empty wine bottles. Flotsam and jetsam decorated the walls alongside letters of goodwill from around the world.

'Ten thousand strangers sent money to the Eigg appeal,' Maggie said, beaming. 'We raised £500,000 in total. Schoolchildren sent pennies and grannies sent their weekly pensions. And then there was the anonymous donation of £1 million from a woman who hasn't even been to the island.'

'Why did she make such a generous contribution?' I asked.

'She felt very strongly that the island should be in the hands of the community, not just sold willy-nilly to the richest person.'

The islanders printed T-shirts and leaflets and launched a website. Contributions arrived with every postbag. The Eigg appeal finally celebrated the successful buyout on 12 June 1977 with the Liberal Democrat leader-to-be Charles Kennedy and the Historiographer Royal Professor Christopher Smout in attendance. The Minister for the Highlands and Islands unveiled a bronze plaque and announced, 'Game, set and match to the islanders!'

The island is now run by the Eigg Heritage Trust, a consortium of islanders and their business partners. 'The only problem is that there is no one else to complain about,' said Maggie, giggling. Until now of course.

'What do you think about the recent press articles?' I asked.

'Nothing but sour grapes,' she replied emphatically. 'We have achieved a great deal and lit the candle for many other community buyouts throughout Scotland.'

I returned to the Land-Rover I had hired for my stay to continue my journey around the island and headed towards the famous cave on the south coast that Maruma had likened to that sacred part of the female anatomy. It was here in 1577 that a party of men from the MacDonald clan carrying a pitchstone rock from the An Sgurr to the Kildonnan graveyard saw the ships of a MacLeod raiding party approaching the island. Believing themselves outnumbered, they dropped the rock, gathered their clansmen and hid in the cave, but a light snowfall allowed the MacLeods to track them down. They lit a brushfire in the cave's mouth and suffocated 398 MacDonalds in a matter of minutes. Two hundred years later, a clergyman visiting the island reported that 'the bones are still pretty fresh, and some of the skulls entire, and the teeth in their sockets.' To this day the pitchstone rock remains where it fell.

The imposing An Sgurr of Eigg, which dominates the island, is a tall column of lava thought to be the biggest volcanic rock mass in the UK. It is said that the world's last pterodactyls roosted here among the 1,500-foot-high cliffs. As I walked below them it was easy to imagine that I was in a prehistoric world, dwarfed by the landscape around me.

I soon found the little loch where, as a small boy, I had spent an afternoon fishing with Schellenberg, oblivious to the mounting ill-feeling towards him amongst the islanders. I recognized the tiny cottage where I had stayed, now sadly abandoned and decaying. I walked up and down the Singing Sands, as the locals call them, just as I had done a quarter of a century before, when I had marched up and down the beach all afternoon willing the sands to sing to me as my nanny had promised they would, but managing to win nothing from them except a few blisters and some tears of frustration.

Walking around this remote corner, images from my childhood flooding back, was strangely nostalgic and quite emotional. I was very happy to be back here and happier still that this dogged and resourceful community had succeeded in wresting their land from the hands of owners who had made their lives troublesome and upset their way of life for so long. Maggie's and the islanders' achievements can only be applauded and it was heartening to know that they had blazed a trail for other remote Scottish communities to follow.

## Gruinard Island

If Eigg and Muck are enchanting and beautiful islands with happy communities and an enviable way of life (at least to me), the same most definitely cannot be said of Gruinard. A little further up Scotland's spectacular western coast, near the fishing port of Ullapool, lies a bleak, treeless little island a mile long and half a mile wide. On the face of it, Gruinard is indistinguishable from hundreds of other islands along this coast, but this one has a sinister and disturbing history. It is better known as Anthrax Island.

For almost fifty years, noticeboards placed around it warned people to 'KEEP OFF'. During that time, strangely suited figures could be seen once every few years wandering across the grass and heather, taking samples of the soil. Anthrax Island was the place where during the Second World War British scientists had exploded a series of biological bombs. The island remains shrouded in much secrecy, but as the years have passed more and more pieces of a mysterious jigsaw have been pieced together to reveal a dark history.

In 1941, Britain bought Gruinard Island from its owner for £1,500 for use by Porton Down, Britain's secret centre for biological research based in Wiltshire, and the isle was christened Base X. As part of the British government's experiments with germ warfare, Gruinard was heavily contaminated with *Bacillus anthracis* (anthrax to you and me), the ancient bacterium that plagues many animals, most commonly sheep and cattle, and which can be fatal to humans unless they are quickly treated with large doses of antibiotics.

In 1942, scientists packed anthrax-laced gruel into a variety of explosive devices and detonated them around sheep confined to boxes. The sixty 'guinea pig' sheep all died within a matter of days and were buried on the island. The key discovery was that anthrax survived the heat of the explosions and was far more lethal than existing chemical weapons. Death from anthrax can happen slowly or instantaneously, depending on how the bacillus finds its way into its host (through the skin, lungs or mouth). If it enters via a lesion on the skin the itchy bumps develop into black ulcers (the word anthrax comes from the Greek for coal), lymph glands swell and death occurs within a few days. If it is inhaled, symptoms resemble a cold until the victim dies of shock. And if the bacterium finds its way into the mouth, the victim suffers fever, nausea, vomiting and, in 95 per cent of cases, death. Experts on biological weapons have suggested that 100 kilograms of anthrax could kill more than three million people.

Gruinard was a curious location for the Porton Down scientists to choose as the test bed for their experiments, as it sits just a few hundred yards from the Scottish mainland. The precautions taken by the scientists were not enough to stop anthrax from spreading to the mainland in 1943. The dead sheep had been dumped in a cave on the island and buried under tons of rock, but one of the carcasses was dislodged during a storm and floated across the bay to the shore, where it infected and killed seven cows, two horses, three cats and more than twenty sheep – and that's just according to government records.

The government – which didn't reveal its germ warfare tests until 1947 – wrongly presumed that the anthrax would die off on its own. In 1979, scientists discovered anthrax spores deep within the soil and for the first time the government erected warning signs along the shoreline. Then in 1981 a group of radical environmentalists calling themselves Dark Harvest claimed to have removed 300 pounds of soil from Gruinard Island and deposited it at the gates of Porton Down and at Blackpool Tower during the Conservative Party conference. The event caused a considerable amount of angst, but the Chemical Defence Establishment played down the risk, just as it had when an Austrian tourist was stranded on the island in 1982.

In 1986 squads of scientists descended on the island's three poisoned acres and, using thirty miles of perforated hosing, decontaminated them by soaking them with a mixture of seawater and 280 tons of formaldehyde to a depth of fifty feet into the ground. The topsoil was removed in containers. In the summer of 1987 the Chemical Defence Establishment announced that the island was now safe for visitors. In 1990 the 'KEEP OFF' signs were removed and the island was returned to its original owners, the Gruinard Estate.

Nasty stuff, and certainly the island is not a place to attract the casual visitor. Unless of course you are a nosy writer, researching a book about offshore Britain, in which case it resembles that fourth brownie to a chocoholic.

'Gruinard?' asked the fisherman incredulously. 'You want to go to Anthrax Island?'

I nodded.

'You know what went on o'er there?' he continued.

'Yes, but apparently it's safe now,' I said, smiling nervously. This wasn't quite the reaction I had been hoping for. I needed more enthusiasm.

He arched his eyebrows and shrugged his shoulders, and we struck a deal. 'I'll drop ye off, but I ain't staying there.'

Minutes later we were 'putt-putting' across Gruinard Bay in his small fishing boat, on our way to one of the world's most notorious islands.

In spite of the official reassurances, I still had reservations about visiting the island. To prove that the clean-up was successful, a flock of sheep had been allowed to graze on it at the behest of an independent watchdog set up by the Ministry of Defence (MOD), and then on 24 April 1990, the then junior Defence Minister Michael Neubert made the half-mile journey from the mainland to declare Gruinard safe. At the same time, however, it was reported that a leading archaeologist was unconvinced by the official position that it was safe to land there. Dr Brian Moffat, archaeological director of an excavation of a medical hospital near Edinburgh, said his team had encountered buried anthrax spores that had survived for hundreds of years. Anthrax occurs naturally in the wild and historically it was not uncommon for animal workers to become infected through skin contact with livestock; indeed, infection by anthrax was once called wool-sorter's disease. Dr Moffat was quoted as saying, 'I would not go walking on Gruinard. If anthrax is still active on Gruinard, there is no reason to suppose it has not survived on more recent sites. It is a very resilient and deadly bacterium.'

'Are you sure about this?' asked the fisherman as we swept across the bay.

To be honest, I wasn't. All I had was a vague assurance from the MOD that the island was 'probably' safe. I had spent the previous week scouring army surplus shops in search of a protective suit and gas mask. Incredibly they were all sold out. All I could get my hands on was a pack of surgical masks from Boots.

I had been slightly reassured by the fact that a Scottish team of scientists had recently produced a vaccine, but I was still wary. It seemed to go against all my instincts of self-preservation. After all, it was only in 2001, when the world was on high alert after September 11, that anthrax had sent terror sweeping through Washington after small parcels of it started turning up in mail rooms in the Senate. As a result journalists from across the world had descended on the tiny, featureless Scottish island, many of them donning biological warfare suits and gas masks as they trudged ashore. Some of the more ridiculous newspaper stories even claimed that Scottish terrorists had supplied Osama Bin Laden with scoops of soil from the island.

As far as I could make out, the main beneficiaries of that media invasion had been the local fishermen, who charged journalists £500 for the short trip to the island. The *Guardian* had mocked the media stampede, joking that the island provided 'a fertile nesting environment for a migrant population of London-based journalists desperate for a new angle on the war against terrorism'. Poking fun at a rival publication, it continued, 'large patches of lime green vegetation struck fear into the heart of a *Daily Telegraph* reporter apparently unacquainted with the concept of moss.' The island had since returned to normality, as had ferry prices – not that there were many takers. One local told me that I was the first person to seek passage to the island since the 2001 invasion.

Sailing across the bay brought home to me just how close the island was to the mainland and its small settlements. It seemed incredible that the MOD deemed it remote enough to carry out such dangerous experiments. I was also struck by how green and lush it looked. For some reason I had half expected a bleak moonscape, not an island alive with vegetation.

'I hope you don't mind getting a wee bit wet,' hollered my ferryman with a wry smile as we neared a small headland with a pebble-strewn beach. 'Are you absolutely sure you want to go ashore?' he bellowed over the diesel engine that was spewing out clouds of black smoke.

'Well, it's now or never,' I said, grimacing, my heart beating, and with a hop and a splash, I landed on Anthrax Island.

'I'll be back at lunchtime,' the boatman called as he disappeared back into the bay.

Suddenly I felt very, very lonely. Not normally

spooked, I found myself rooted to the spot, unable to move. There was an eerie silence. Eventually I gained control of my jelly-like legs and made my way up the beach towards the interior of the island, pebbles crunching under my feet. I haven't come this far to give up now, I thought as I dipped into my bag for a white face mask and set off like a Japanese tourist with a cold.

When the anthrax experiments on Gruinard first became public knowledge after the war, the official explanation was that anthrax spores were being cultivated in an attempt to create a cure should the Germans ever deploy anthrax as a weapon. That, however, was only part truth. The other reason for the experiments only came to light ten years ago when documents emerged from the Public Records Office about a wartime venture codenamed Operation Vegetarian. The plan was simple: inject millions of small cattle cakes with anthrax and then drop them from bombers over the large beef and dairy herds that grazed over north Germany. The animals would eat the cakes and die, leaving Germany with an inevitable food shortage, and of course should anthrax cross into humans – well, that would be Hitler's problem.

The man in charge was Dr Paul Fildes from Porton Down's biology department. A number of practical difficulties needed to be overcome. First he needed to source five million linseed cattle cakes, and then equipment for injecting each cake with the spores had to be designed. Special containers would have to be made to carry the contaminated cakes, and bombers had to be adapted to be able to deposit payloads of cow food rather than bombs. Tenders went out to various manufacturers, and the contract eventually went to Messrs J. & E. Atkinson of 24 Old Bond Street in central London, perfumers and toilet soap manufacturers, and suppliers of soaps and unguents to the royal family.

Atkinson calculated that it could produce 180,000 to 250,000 lozenge-shaped cattle cakes in a forty-four-hour week. Each cake would be an inch in diameter and weigh ten grams. It promised the five million cakes as long as its workers were exempted from military service. Cakes sorted, Fildes now had the small problem of injecting them with the anthrax bacillus. Thirteen women were recruited from several soap-making firms, shipped down to Porton Down, sworn to secrecy and tasked with injecting five million cow cakes with anthrax.

In the meantime Lancaster, Halifax and Stirling bombers were fitted with wooden trays from which they could deposit their tasty payloads. Perhaps unsurprisingly, the RAF weren't overjoyed at the prospect of becoming aerial farmers, or indeed of contracting anthrax themselves. Officers assured crews that their thick leather clothing would resist the spores, and by the beginning of 1944 Operation Vegetarian was ready to go. Fildes set out his strategy to colleagues: 'The object of this contemplated operation is to infect and kill the largest number of cattle possible in enemy territory in a single effort by means of small, infected cattle cakes dropped from the air.' He estimated that a Lancaster returning from a raid on Berlin would be able to scatter 4,000 cow cakes, and that 100 bombers could therefore deliver nearly half a million cakes. But by the time Fildes's operation was ready to go in the summer of 1944 the Normandy landings had taken place and Allied armies were heading through northern France. It seemed that the war against Nazi Germany wasn't going to be won with cow cakes, and the five million anthrax lozenges were incinerated at Porton's high-temperature furnace.

As I trudged across the thick machair and heather I was struck by the number of rabbits. According to locals, Gruinard is home to two golden eagles, for which the rabbits make rich pickings. I couldn't help wondering how deep their holes went. What if the rabbits had reached below the decontaminated line? A small burn babbled with peaty water next to a dilapidated crofter's cottage, lobster pots stacked on the wall. An application had apparently been made to the Highland council in the mid-nineties requesting planning permission to build a home on the island. The application was never pursued.

Alone I trudged on, my boots squelching in the boggy soil. Sheep grazed happily, and contrary to expectations they didn't all have five legs and neon pink coats. In fact there was nothing to distinguish the island from thousands of others, I thought, as I returned to the beach, where seals swam contentedly in the shallows. I sat down and removed my face mask, and poured myself a cup of tea from the Thermos kindly packed for me by my worried B&B landlady. 'In case you get marooned,' she said as she handed me an emergency knapsack. How thoroughly British, I thought, as I sipped my tea on Anthrax Island.

I breathed a sigh of relief as I spotted the little fishing boat once again, crossing the bay. It had been a distinctly unremarkable visit to a remarkable island. I had spent only a few hours there. I can't say I was sad to leave Gruinard. In fact I was immensely relieved. The island may no longer bear the physical scars of its bleak history but its air is thick with it. The island had continued to have an important role even after the war, when it became a refuelling point for nuclear submarines and a training ground for NATO troops. How the servicemen and women must have hated those deployments! It remains a pariah. It has even been highlighted as a possible future dumping ground for nuclear waste. For this small island it seems that the future's not so much bright as positively luminous.

# St Kilda

Norman Gillies fixed his eyes on the horizon like a bird of prey viewing its quarry. The wind snapped at his jacket, tousling his hair and watering his eyes. A rogue wave crashed over the bow of the boat, soaking us with its salty spray. Unblinking, his eyes flicked across the vast expanse of the ocean, scouring the deserted water. His tears began to well up more obviously as a small blur of greyish black slowly came into focus in the distance. We were nearing one of the most remarkable set of islands in the British Isles, an ancient place that has stood uninhabited for the last seventy-five years. I was heading there in the company of one of the survivors of the last community to live there. St Kilda, the collective name for the four main islands Hirta, Dun, Soay and Boreray, rises proudly out of the Atlantic, its imposing cliffs battered by wave and wind, a timeless memorial to an ancient, lost society that finally folded, after several thousand years, in the face of advancing civilization. When the naturalist James Fisher visited St Kilda in 1947, he lamented that he and every future visitor would be haunted and forever 'tantalized at the impossibility of describing it to those who have not yet seen it'.

The human story of St Kilda began in the Bronze

Age and ended in 1930 with its dramatic, heartbreaking evacuation. In the 1830s the Reverend Neil Mackenzie found what were almost certainly the remains of pre-historic burial chambers in Village Bay on Hirta, the only realistically habitable site on these storm-battered, craggy outposts. In 1844 an Iron Age house dating from about two thousand years ago was discovered, comprising a number of passages with cells branching off them. Two stone crosses from the early Christian period have also been found.

The first owners of St Kilda, according to Celtic legend, won the islands in a boat race. The MacLeods of Harris and the MacDonalds of Uist (two islands of the Outer Hebrides) both claimed the island as their own. To settle the dispute the contenders agreed to a boat race across the fifty miles of open sea to St Kilda. It was stipulated that the two boats were to be the same size and type, crewed by an equal number of men. The first person to touch land on Hirta would be declared winner and secure possession of the island for his clan and chief. The race was close, but as the two boats rounded the outer reaches of Village Bay the MacDonald crew drew a few lengths ahead. The exhausted MacLeods had all but given up hope when one young crew member, Coll MacLeod, threw down his oar, drew his sword and severed his left hand and, with a triumphant cry, hurled it over the heads of the MacDonalds. The hand sailed through the air and landed on a rocky promontory, thereby winning the race and its coveted prize for the MacLeods. To this day, the red hand in the MacLeod coat of arms is said to commemorate Coll's gallant action.

St Kilda's geography is as dramatic as its legendary roots and it made for an awesome sight as Norman and I approached that morning aboard *Orca*, a converted fishing boat. It nestles in the ferocious north Atlantic almost fifty miles west of the Hebrides, and it seems incredible that anyone could ever have lived here at all. The sea cliffs of Hirta soar 1,000 feet into the air, pounded by some of the heaviest seas on earth. There is no protection here. Not even a tree.

I was overwhelmed by the scale of the near-vertical cliffs as we pulled into Village Bay. Staring at their vertiginous walls was dizzying; I felt like a first-time tourist to New York, dazzled by the skyscrapers. The cliffs are reputed to be the highest in Europe and they dwarfed our little boat. Banks of sea mist rolled down from the slopes of the island's interior, occasionally but only briefly disappearing altogether to reveal the 1,400-foot peak of Conachair in a burst of bright sunlight, before it was once again enveloped in woolly cloud. It looked Jurassic, I thought, as we pulled into the only reasonably safe anchorage and settled into its calm waters.

'There's my house,' said Norman, pointing to a tiny row of houses strung out along the valley like a stone necklace, his face now overcome with a grin to match the scale of the landscape. Norman stood out from the rest of the visitors aboard the *Orca* with his shirt and tie and carefully polished shoes – a funny contrast next to my sailing jacket and walking boots. On first acquaintance Norman is like any other octogenarian with his thick-rimmed glasses and balding head, but he was far more sprightly and energetic than most people of his age I have come across, bouncing around the boat like a mountain goat in anticipation of landing on his homeland.

Being St Kildan, Norman ought to have a strong Gaelic burr, but his accent is an intriguing mix of Scottish and 'Estuary' English that betrays the island's history and his personal odyssey. For the past fifty-seven years Norman has lived in Ipswich in a house called St Kilda, a world away from the Atlantic out-crop that was home to scores of his ancestors. When St Kilda was evacuated in 1930 and the last families of an ancient community, carrying their furniture on their backs, left their homes for ever, it was the final chapter in a way of life that had remained largely unchanged for nearly two millennia. Norman is one of only three remaining evacuees (his two surviving relatives from St Kilda live in Scotland, one on the Black Isle and one at Greenock). Norman was just five when he was evacuated. He had been back once before; this time he was heading 'home' with his son John (who gave up his job as a printer in 1987 and spent the next few months tracking down the remaining St Kildans).

A small dinghy was launched from the *Orca* and we were ferried to the storm-damaged pier. Until this jetty was built in 1901, all people and goods had to be landed precariously on the rocks. St Kilda has always been a prisoner of her location, and often the island is cut off for many months at a time.

In the 1950s the military set up an ugly base by the shore and vast radar towers on the summit of Hirta. It is a tragedy that an island of such outstanding beauty should have been scarred by such an ugly installation. Although the army has gone, civilian contractors still work here from time to time, tracking practice missiles fired from Benbecula on North Uist. Today the island is owned by the National Trust for Scotland, which runs work parties that help the island rangers restore the row of houses. There are also scientists at work for Scottish Natural Heritage surveying the island's famous Soay sheep. The island has dual World Heritage status, for its natural and cultural significance.

Nobody lives here permanently, but following the launch of a daily boat service from Harris in the Western Isles during the summer months the island receives up to thirty new islanders each day. As we poured ashore, and the day-trippers began setting off in all directions to explore, I felt a dim sense of how life must have been for the inhabitants when the Victorian cruise ships offloaded their cargoes of tourists to ogle at the natives, paying them to have their pictures taken with them and swapping the islanders' valuable tweed for much-sought-after fruit and other 'exotic' goods. One story tells of a St Kildan swapping her entire year's worth of tweed for an orange. It is generally held that early tourism corrupted the islanders. By the end of the nineteenth century the St Kildans had also developed a significant dependency on charity. Once the islanders even burnt a new boat given to them as a gift because it wasn't good enough, in full expectation that another would be sent shortly.

There was a palpable sense of loss here, amid the ruined houses and the hundreds of stone cleits – the tiny stone huts built to store and dry out the thousands of birds the St Kildans hunted each year. Situated in the most sheltered part of Village Bay and protected from the winds by the heavy shoulder of Conachair, the small settlement or village on Hirta consists of around thirty black houses (typical crofters' dwellings), strung out for half a mile along the Street. Built of stone, they all had low doors and small windows and were strongly constructed to withstand the onslaught of the weather from which there was rarely any respite.

Walking with Norman on the island that morning was like stepping back in time. He was walking, talking history – the last dodo of St Kilda. We entered one of the houses, the inside of which was divided into two rooms by a moveable stone partition called a fallon. In winter and spring one side was occupied by the family and the other by the family cow. 'This was my house – number ten,' announced Norman proudly. 'There used to be a dyke in front of it.' Most of the house had been eaten away by the weather and all that remained was the skeletal walls with their windows, doorframe and fireplace. 'This was where I slept,' he said, pointing to a sheep munching at a tuft of grass.

'And this was where we ate,' he added, pointing to another barren area a few yards away. 'When I was out playing at the other end of the village, my mother would stand out on the street and call out, "*Tormod iain*, time to come to dinner." I only spoke Gaelic then. I didn't learn English until I went to school in Lochaline in Argyll.' The passage of time has erased most of Norman's childhood memories, although he still recalls filing into the tiny little church with the rest of the islanders. 'We went twice on Sunday and there was no work. It was a rest day.' In 1697 a visitor called Martin Martin recorded in his book *A Late Voyage to St Kilda* that the islanders loved their music and games, but by the late nineteenth century the islanders' commitment to the Free Church of Scotland had led to a less fun-filled life.

Norman's most enduring memory is the day St Kilda died as a community, 29 August 1930. He described how it came about. 'My mother was pregnant and took ill with appendicitis,' he recalled. 'First of all they had to get a message out with a fishing trawler that there was someone ill. The lighthouse ship came but the weather was so rough that they couldn't get a boat out from the shore. The boat came back a few days later and she was taken to Stobhill Hospital in Glasgow, but it was too late and she died. My little sister died a few days after she had been born.' The tears welled up in Norman's eyes once again. 'My most precious memory is of my mother being rowed out to the lighthouse ship, with her shawl over her head, and waving to me on the shore. That is a real treasure I will remember all my life.'

Communication was always a problem for St Kilda. The only way islanders could communicate with the outside world in times of distress was either by lighting a bonfire on top of Conachair in the hope that it would be spotted by a passing ship or by sending a message via a St Kilda mailboat. This craft consisted of a piece of wood, the size and shape of a toy boat, hollowed out in the middle so that it could hold a small bottle or tin, which contained a letter, instructions for the finder to post it and a penny for a stamp. The bottle was waterproofed with grease and battened down under a little wooden hatch bearing the inscription 'Please open', burnt in hot wire. A float made of an inflated sheep's bladder with a small red flag tied to its mast was attached to the hull of the mailboat, which was then launched when the wind was in the north-west. Mailboats often turned up on the west coast of Scotland or sometimes even in Norway.

I asked Norman if he ever visited the grave of his mother and sister. 'No,' he replied sadly. 'I wouldn't know where to find her.'

The death of Mary Gillies in January 1930 was the island's 'tipping point', a small but tragic incident that brought the whole of the island society crashing down. In May twenty islanders signed a petition to the government requesting evacuation. In truth, though, St Kilda had begun to die years earlier. Organized religion, a dependency culture based on charity and tourism, emigration and economic changes all eroded the relationship the islanders had with each other and with nature. Why St Kilda came to be evacuated has been the subject of much debate over the years. Some point to the effect of the persuasive missionaries on the island. One story in particular illustrates the problem vividly. In 1877 there was a severe food shortage on the island and many islanders had begun to suffer from malnutrition. One Saturday in early May a ship called HMS *Flint* arrived off Village Bay, carrying supplies for the islanders. That evening the *Flint* cast anchor and the captain came ashore. He was met by the minister from the church, who informed him that 'as the people must prepare for the devotion of the morrow, they could not think of encroaching on the Sabbath by working at landing any goods'. So the *Flint* was sent on her way, leaving the islanders bereft of their much-needed supplies.

Life on St Kilda was always difficult. Along with traditional crofting activities such as peat cutting, cattle grazing, fishing, spinning and weaving, the islanders eked out a living working the islands' treacherous cliffs for the seabirds. St Kilda has always been a major seabird breeding location, and the islanders hunted gannets, fulmars and puffins for eggs, meat, feathers and oil, some of which they consumed themselves, while the rest went to pay the rent to St Kilda's land-lord, latterly the MacLeods of Skye, whose steward came annually to the island to collect it. Nothing was wasted – even the entrails were used for manure. A typical yearly harvest was 5,000 gannets, 20,000 puffins and about 10,000 fulmars. The birds were taken during the breeding season, which lasted from about March to September, caught by hand, or with a fowling rod or a snare. Hunting them was a difficult and dangerous business. It involved an open boat trip, which could only be undertaken in calm weather, across four miles of Atlantic Ocean to Boreray, the largest gannetry in the world. Ganneting had to be done on a moonless night. Two men with ropes around their waists climbed up the cliff face until they reached the ledges where the gannets slept. They then had to identify the sentinel bird that kept watch while the other birds slept, rather like a night watchman. Once they had identified it, they would sneak up and dispatch it. The expedition depended on the success or failure of this exercise. Once the sentinel had been killed, it was comparatively easy to catch the sleeping gannets, wringing their necks and throwing them into the boat below. In the seventeenth century, St Kildans were catching 22,000 gannets a year. Once back on Hirta, they would pluck the birds' feathers, extract the oil and salt and store the carcasses.

While bird harvesting may be a thing of the past on St Kilda, it is still an integral part of life for another remote Scottish isle, Sula Sgeir, an uninhabited outcrop some forty miles north-west of the island of Lewis. It is just half a mile long and, like St Kilda, ringed with precipitous cliffs. For two weeks each year, the men of

Ness on Lewis undertake the last surviving bird cull in the UK. When the Wildlife and Countryside Act was passed in 1981, special dispensation was written into it to allow the cull of 2,000 guga on the remote island by the ten families that are the men of Ness. In August each year, the men leave the comfort of their insulated homes and live in stone beehive houses – sparse bothies that they share with earwigs and the wind. They rise early and work until midnight, trapping, gutting and salting the birds before packing them into barrels. Places on the hunt are prized and outsiders are rarely invited. Perhaps unsurprisingly, it has attracted a fair amount of criticism in recent years from campaigners who argue that there is no place for the culling of gugas in a modern civilized society.

Dods Macfarlane, veteran of thirty-one hunts and unofficial spokesperson for the hunt, argues that while each nation enjoys its own delicacy, the guga is integral to the men of Ness's traditions and heritage, as the gannet was for the St Kildans. And what is the difference between hunting the guga and hunting pheasant, partridge and grouse, or indeed feasting on turkey at Christmas? I had heard that when the boat returns to Lewis, the queue of locals waiting to purchase the delicacy stretches for more than two miles and that each bird sells for up to £20 – certainly a higher premium than the twenty pence a grouse fetches. I had been desperate to accompany the men on this age-old hunt but had been politely let down by mild-mannered Dods. 'I'd love to invite you,' he said, 'but not in this climate. We don't want to stir up any more controversy than we already have.'

The St Kildans spun and wove wool from the world-famous Soay sheep who, like the gannets, live on the precipitous cliffs. According to legend, a Viking by the name of Callum brought the first sheep to Soay, although it is argued that they might have arrived much earlier with Neolithic settlers. Soay sheep are a primitive breed descended from the mouflon, a small wild sheep found in mountainous parts of Sardinia, Corsica, Cyprus and the Middle East. St Kilda is the Soay sheep's last native habitat in the world. In his *History of the Scottish People*, published in 1527, Hector Boece describes Soay sheep as 'wild beasts not very different from sheep. The hair is long and drab neither like the wool of sheep nor goat.' The sheep are mainly a deep chocolate-brown colour, and they have long legs that enable them to run fast and give them the agility necessary for St Kilda's steep, mountainous cliffs. From the cliffs' edge you can see them grazing the grass on narrow ledges hundreds of feet below in seemingly totally inaccessible places. When I visited St Kilda I came across a team of four Ph.D. students studying the world's 'purest' Soay sheep, untouched by humans for seventy-five years.

The vertical cliffs not only characterized the island but determined some of the physical attributes of the islanders. While a sixth toe seems to have been a figment of Victorian imagination, they certainly had unusually strong and agile feet with which to scale the vast cliffs. (Norman volunteered his bare feet for a toe count. There were indeed ten toes.) Their toes acquired an almost prehensile capacity and were set wider apart than those on a normal foot. The instep was thicker, too, and the ankle more muscular, with a heavier bone structure better suited to barefoot climbing.

The cliffs were also part of a rite of passage. Before a young St Kildan could get married he had to demonstrate that he was capable of supporting a wife and a family by giving a display of courage and skill as a cragsman. Near the top of Ruaival on the south-west coast of Hirta there is a natural arch in the stone known as the Mistress Stone. It is formed by a great slab of granite, which juts out from a pinnacle of cliff. The surface of the stone is flat, overlooking a sheer drop of 250 feet to the raging sea. Observed by friends and future fiancée, he had to stand on the lip of the stone and, balancing upon the heel of one foot, bend forward and grasp the other foot with both hands. He had to hold this position, looking down at the rocks and surf below him, until his friends decided he had proved himself.

There are myths aplenty on St Kilda but some of the true stories are equally intriguing. My favourite is that of Lady Grange. In 1715 James Erskine, Lord Grange, married Rachel Chiesly. The marriage soon deteriorated and the new Lady Grange decided to take revenge on her husband. Her weapon was a document that proved that her husband was involved in the Jacobite uprising. Word reached Lord Grange, who realized the potential danger and had his estranged wife kidnapped and sent to St Kilda. She lived in a two-room cottage some way below the village in the bay area. As no one on the island spoke English and there were no books, she took to sleeping by day and living by night. She drank as heavily as her meagre supplies of whisky would permit and every night wandered the shoreline, wailing and bemoaning her fate. Slowly madness took over her frail body. Throughout her seven years on the island she made repeated attempts to contact her friends on the mainland. Eventually she managed to hide a letter in a bale of yarn, which was collected as part of the rent by a steward and taken to Inverness. The letter was dated 20 January 1738 but did not reach the Solicitor General, to whom it was addressed, until the winter of 1740. It told of the poor woman's ordeal and ended with the plea: 'When this comes to you, if you hear I'm alive do me justice and relieve me, I beg you make all haste but if you hear I'm dead do what you think right before God.'

The letter created a great stir in Edinburgh, and eventually after legal attempts by Lord Grange to block a rescue, a ship of armed men was dispatched to St Kilda, but by the time they arrived Lady Grange was gone. She had been removed to Skye, where she spent the last twelve years of her life in captivity and eventually died insane.

For the seven years Lady Grange was a castaway on St Kilda she was accorded exemplary hospitality by the islanders. A young St Kildan waited on her and the islanders made sure she always had all the food and warm clothes she needed. Nevertheless, her exile must have been an awful punishment for someone used to the comforts and security of a gentrified life.

The St Kildans' attitude towards strangers and visitors in general was complicated by one factor common to all remote islands around the world: infection and epidemic. Until the middle of the nineteenth century most islanders would always react with alarm to the arrival of strangers and disappear into the hills. The St Kilda 'cold' has many parallels to that of Tristan da Cunha, Pitcairn and Fair Isle, and was the cause of many deaths because of the islanders' lack of immunity. Perhaps the worst case was one in the eighteenth century. In 1723 an old man from St Kilda went on a visit to Harris, caught smallpox and died there. The following year some of his relatives went over to Harris to fetch his belongings. Among these were the old man's clothes, which were still infectious. When the party returned to St Kilda, smallpox decimated the community and almost everyone died. Some of the men, however, had gone on a fowling expedition to Boreray before the epidemic struck, and because there were not enough able-bodied men on Hirta to launch the boat and fetch them, they remained there and survived on the tiny pinnacle of rock for nearly six months until the next steward's visit. When they returned to the main island they found that only four adults and twenty-six children had survived. The community had to begin life again.

Walking the island was eerie. I could almost hear the islanders of old in the wind as it whistled up the craggy cliffs and funnelled down the valleys littered with cleits which gave the island, from a distance, the appearance of a giant graveyard. When I returned to the Street, Norman was holding court, the volunteers hanging on to his every word. His memory augmented by what he had digested from the many books and articles he has read about the island, he was an umbilical cord linking them to the island's past. We munched our packed lunches and drank tea provided by the work parties as the guest of honour reminisced. Workers doffed their caps at this St Kildan celebrity as they headed into the hills to count sheep. Norman was enjoying the opportunity to bring the land of his forefathers back to life.

'That isn't Neil Ferguson! It's Donald MacQueen!' he snapped, as he looked at a black-and-white postcard of two islanders on sale in the little gift shop. In the small museum he marvelled at the photographs of friends and relatives: 'I remember that one being taken . . . that was my Uncle John.' Norman pointed to a dour-looking face heavily lined by the weather. 'He drowned,' he said matter-of-factly. There was no tradition of swimming on St Kilda – presumably because the waters were too dangerous for the islanders to learn – and drownings were common among the men who regularly rowed the four miles to Boreray. 'No St Kildan died in his bed,' recorded one visitor to the island. Norman's son John is named after the uncle

– another link to the past. St Kilda hangs on to her history by her fingernails, just as the gannet hunters of old held on to the steepling cliffs they scaled.

Of the two dozen photographs in the museum, I was struck by the fact that there was not one of an islander with a smile. All the faces seemed contorted and lined with hardship. I was particularly taken by a photograph of a woman and her young child. Her head was covered in one of the islanders' distinctive scarves with its white frill. She wore a long heavy skirt and the customary grimace. 'That's me,' said Norman, a broad smile breaking across his face. 'Wasn't my mother beautiful?' I shivered. The history of the place was almost tangible to me in that moment of Norman's revelation. Noble savages or lost utopians, the people of St Kilda continue to thrill and to inspire.

Near the ocean edge, nestled among the military buildings, is the island's famous Puff Inn, the western-most pub in the UK. The ceiling and walls are plastered with flags and pennants from a thousand passing ships, and a vast banner above the bar announces 'The Great British Rockall Expedition', left there in 1955. A television showing a football match blared from above the bar. I asked Norman if there had been a pub when he lived there and he arched his eyebrows. There was no time for fun or living it up on St Kilda. Life was about surviving from one day to the next.

The Puff Inn is the social hub of the island for visitors and a place where personnel from the missile tracking centre and work parties involved in managing the island's wildlife can socialize. There is, however, little love between the two bodies. 'One party is here to destroy, the other to preserve,' explained a National Trust volunteer, sinking a can of Guinness.

On that fateful summer's day in 1930, HMS *Harebell*, the SS *Hebrides* and the *Dunara Castle* anchored in Village Bay to evacuate the St Kildans. The sheep and cattle were loaded on to the ships and the dogs drowned in the sea. A bible was left on the lectern in the church. At 7.00 a.m. Norman and the other islanders boarded the *Harebell* and bid farewell to their home. St Kilda was left empty for the first time in nearly a thousand years. Tears streamed down young Norman's face as the *Harebell* steamed out of Village Bay, his home and his community getting smaller and smaller until St Kilda was all but a smudge on the horizon. And then it was gone.

When Norman and his fellow islanders were evacuated to Argyll, incredibly they were given jobs with the Forestry Commission. For many this was the first time they had ever seen a tree. According to Norman, all the trees planted by his fellow islanders have matured now and been felled. The links to St Kilda's past are disappearing fast. 'In Lochaline we were treated very well,' he recalled. 'They didn't look on us any differently from anyone else, although when we came off HMS *Harebell* from St Kilda they were expecting people from outer space.' Dozens of cameras eagerly awaited their arrival and the pictures made front pages across the UK.

Seventy-five years later, Norman boarded the *Orca* anchored off the bay and retraced the *Harebell*'s course as we headed off into the Atlantic wind. Once again tears streamed down his cheeks. He had completed his last pilgrimage and it was time to leave St Kilda to the Atlantic and the gannets once and for all. He stood at the stern of the boat, transfixed as the cathedral-like rocks disappeared into the surf of the horizon, his arms held aloft, waving goodbye to the ghost of St Kilda.

# Shetland

'It's *Shetland*, not the Shetlands,' the hairy Viking corrected me. 'You're on mainland, Shetland, singular,' he continued with all the patience of someone who has had to explain this a thousand times to ignorant outsiders. Was he going to pillage us? I had never met a grumpy Viking before, and certainly not one in a cowhorn helmet. It was bleak and wintry outside as my bearded friend directed me to the car park. I wasn't alone, by any means, having flown in to the island's tiny airport and been escorted to the principal town, Lerwick, with half the world's media. Correspondents from Reuters, Associated Press, the Press Association, the BBC, ABC and CBC flooded off the plane, all neatly wrapped up in their 'just bought from the airport' Millets jackets, clutching their battered silver cases, notepads and sound recorders. We had all come to this far-flung archipelago in the north Atlantic for one of the largest fire festivals in the world, Up Helly Aa. And the Vikings were about to get a whole lot grumpier.

Shetland is the most northerly of all Britain's islands and it certainly felt like a foreign land, part Scottish, part Scandinavian but wholly *Shetland*. Singular. This group of 100 islands, 15 of which are inhabited, sits

600 miles from London and just 400 miles south of the Arctic Circle, which is as far north as St Petersburg in Russia and Anchorage in Alaska. The Norwegian city of Bergen is 225 miles east, closer than Aberdeen. Even freezing Labrador in Canada lies further to the south, but Shetland enjoys an extraordinarily mild climate. This is because warm currents from the Atlantic flow by and mix with the colder waters of the Norwegian Sea to create a temperate maritime climate, which changes very little between the seasons.

The weather here, though, comes with a double warning. It is both extremely windy and crazily changeable. One minute the islands are bathed in pure blue sky without a breath of wind to be felt; the next they can be blown into a frenzy by a gale-force bluster. In 1962, a gust of 177 miles per hour was recorded here, the strongest ever in Britain, but it never made the official books because the wind carried away the measuring system and there was only the good word of the meteorologist to vouch for it. One of the first tell-tale signs of the powerful winds is that there is nothing tall on the islands. For a start, there are no trees anywhere on this remarkable archipelago, described once as a 'drowned jigsaw', which spans over a hundred miles from Fair Isle in the south to the fantastically named Muckle Flugga lighthouse in the north. Everything is squat. Even the Shetland ponies, cattle and pigs have been reduced to toy size by nature so that they don't get blown into the sea.

Shetland proved a pleasant stopover and refuelling

point for marauding Vikings on their way to beat up as much of the known world as they could find. They invaded and settled the islands way back in the eighth century, and their blood still runs thick today, as we quickly discovered after stepping off the plane. Shetland came relatively late into the Scottish fold (in 1469, to be exact) and the Norse have left an indelible mark on the traditions and psyche of the people. Even today, local men take more pleasure in dressing up in the Viking kirtle than the Scottish kilt. With their thickset bodies and bushy beards they even look Viking.

Many Norse traditions and festivals became absorbed into Shetland culture, Up Helly Aa being the most important. As a writer I had been summoned to a 'press meeting', at which the event's strict code of conduct for non-islanders would be outlined. A burly Shetland islander with a gruff voice and thick beard was sitting on a throne. This was the Guizer Jarl, this year's leader of the festival (or Peter Fraser when he's in his day clothes) who leads the 'Guizer squad', made up of a team of men of his own choosing. The honour of being Guizer Jarl is awarded once in a lifetime and our Guizer Jarl was clearly keen to stamp his authority on the proceedings. He stood to address the assembled hacks, announcing, or rather, barking, 'You can't be anywhere near the lighting-up procession, you can't walk with the procession, and you can't come anywhere near the burning site afterwards. Oh, and you can't attend the hall dances afterwards. They're private.' So, in summary, we weren't allowed to go anywhere. We

Vikings abound at Shetland's annual Up Helly Aa celebrations

The gruff Guizer Jarl

The red cliffs of Heligoland

Heligoland's houses are protected from landslides by reinforced steel cages

A Heligoland signpost

Nearby Sandy Island, a haven for hundreds of seals

HRH Prince Michael of Sealand, with his principality in the background

The only way one can enter Sealand

The *Eda Trandsen*

My flag!

The crew aboard *Eda*, on the way to capture Rockall, my island idyll

Circling the treacherous waters around Rockall

More people have been to space than to Rockall

might as well get back on the plane and go home, and good riddance to us. Viking foreign policy had changed little down the years, it seemed.

There was bemusement around the room as Japanese crews and German photographers scratched their heads. The island's tourism representatives shuffled uneasily. Not the PR they had anticipated. The Shetlanders, after all, prided themselves on the warmth of their hospitality. The festival had already gained a reputation for the harsh regime of its organizers. I had heard that Konnie Huq, the *Blue Peter* presenter, had fled the islands in tears after her run-in with the Viking organizers while filming the festivities the previous year.

'So where *are* we allowed?' whimpered one of the photographers.

The Viking shrugged. 'I don't want any of you "soothmoothers" in there. This is a local event for local people,' he snarled. A soothmoother is a derogatory Shetland word for an outsider, someone who has arrived by the southern mouth of Lerwick harbour. Considering one-third of Shetland's 23,000 inhabitants are soothmoothers, and fewer than half the population could lay claim to having three or more grandparents born there, the Viking's strict immigration policy seemed a little hard to stomach.

What a terribly useful meeting that was, I thought, as we walked out of our lecture from the bearded one. I now understood less than I had before the meeting. I had actually managed to delearn. It was late January and already the lunchtime sky was streaked black. A bitter wind funnelled off the water and through the narrow streets. Shop signs swung and creaked in the breeze. Eerie shadows danced across the walls of the deserted town. In June, the sun can shine for twenty hours, setting for just a few hours of twilight, but in mid-winter there are barely five hours of daylight. Winter is long here, and the blues are common.

Lerwick (population 8,000) grew in importance relatively late in Shetland's history. Its early prosperity was built on the shining shoals of herring around Shetland's 900-mile coastline and latterly the oil boom of the 1970s, when oil flowed from Sullom Voe, Europe's largest oil terminal. The discovery of oil – and gas – in the North Sea stemmed the island's gradual depopulation and reinvigorated its flagging economy. That economy remains dominated by fish and oil today but it has become increasingly diverse, with tourism, livestock rearing, quarrying, knitwear and arts and crafts all making significant contributions to the islands' coffers. For the past twenty-five years, the Shetlanders have enjoyed levels of prosperity and standards of living almost on a par with those of their Scandinavian cousins across the water. Like them, they have managed to do so without spoiling the stunning natural environment in which they live and work. Shetland has one of the best-educated workforces in the British Isles, all employed in a variety of different businesses, including computers, wind farming and tidal power, marine engineering, boat building . . .

There is even a strawberry farmer, on the same latitude as southern Greenland.

The huge influx of visitors for the festivities meant that all the best hotels and guesthouses in Lerwick were full to capacity and I was forced to find lodgings in a small B&B on the outskirts of town. Protected in the lee of a cliff, and with sweeping views down the valley to the great north Atlantic beyond, it had been described as 'basic but comfy' in my guidebook, though it was more basic than comfy, with a bed barely wide enough for a pencil and un-turn-off-able heating and unopenable windows that turned the room into a sauna. An old Tourist Board poster of a Viking stared down at me from the wall. They're everywhere, I thought, as I turned the frame around to face the other way. After our experiences that morning I couldn't help but wish that the Picts, who lived here originally back in the eighth century, had somehow managed to fight off the bearded, helmeted ones in their longboats.

It is incredible to think that the Picts and their forebears had inhabited the islands for more than five thousand years before the Vikings rolled ashore, butchering and setting light to everything that stood in their way. (Norwegian legend, incidentally, has it that the Picts were 'scarcely taller than pygmies'. Perhaps it was the wind that made them that way, just as it had the livestock.) The Vikings, known throughout the world by unflattering nicknames such as Stinging Hornets and Sons of Death, had the same roots as the Teutonic tribes that overran most of Europe and contributed to the destruction of the Roman Empire. There is little remaining evidence of the Pict civilization today. The last Pict king, Eoganan, was killed in a battle in AD 839 and the rest of his subjects were wiped out by the marauding Vikings, who went on to establish themselves on the islands, even though it wasn't until the tenth century, when Harald Fairhair, King of Norway, landed on the island of Unst, that the Norsemen officially took control. One of their legacies is one of the world's most popular horses, the Shetland. The Norsemen bred their Dole ponies with the islands' native Equids to produce the diminutive Shetland, which was then domesticated to carry peat and haul seaweed. It is reputed to be the strongest horse relative to its size.

In 1469 the Shetland islands were given to Scotland as part of a wedding dowry for the marriage of James III and Princess Margaret of Denmark. Earl Robert Stewart, a seventeenth-century tyrant who cared little for the islands and ruled its people harshly, compounded the islanders' sense of detachment from mainland Scotland. Today, islanders consider themselves not Scots or Scottish but Shetlandic. Indeed, even today Shetland's Norse heritage is very much in evidence. Most place names are Norse and much of the Shetland dialect, still in use today, is derived from the old Norse language.

Up Helly Aa rose like a phoenix from the ashes of the ancient festival of Yule. The Vikings held Yule,. the most lengthy of all the pagan feasts, lasting three weeks, to celebrate the rebirth of the sun, but with the coming of Christianity, it became Christmas. The twenty-fourth night after Christmas was called Up Helly Aa, or 'the up-ending' of the holy days of the Christmas period. Perhaps in a nod to their pagan past the Norsemen looked forward to Up Helly Aa as the beginning of an almighty binge and carousal. Held in Lerwick on the last Tuesday in January, Up Helly Aa is like a sub-Arctic Mardi Gras. However, while there is evidence that people in rural Shetland celebrated the fire on Up Helly night for hundreds of years, the celebration in today's form did not start until the end of the Napoleonic Wars, when the young soldiers and sailors of Lerwick returned home with a hunger and passion for firepower.

A visiting Methodist minister described the event in his diary in 1824: the 'whole town was in uproar, from twelve o'clock last night until late this night blowing the horns, beating the drums, tinkling old kettles and firing guns, shouting, bawling, fiddling, fifing, drinking, fighting. This was the state of the town all the night.' He could, I suppose, have been describing a Friday night in any market town or city centre across Britain today. By the 1840s the celebration had become even more riotous with the introduction of flaming tar barrels. These were hauled through Lerwick's narrow streets, like molten missiles, burning anyone and anything in their path. The revellers would then leave the burning barrel on the doorstep of that year's least popular islander. The main street of Lerwick in the mid-nineteenth century was extremely narrow, and rival groups of tar barrellers frequently clashed in the middle, forcing the town council to appoint special constables to control the raucous chaos that inevitably ensued.

In the nineteenth century the festival became even more riotous and special constables were called in from the mainland to curb trigger-happy drunks firing their guns into the air. At the same time the council and the more sober town residents stepped in to try to 'civilize' the proceedings and hit upon the idea of substituting the barrel with something more akin to their Norse ancestry, a Viking longboat. Thus was born the Up Helly Aa we know today.

The festival has grown considerably through the years, as had the worldwide media interest in it. While the coverage has undoubtedly helped the island's tourism industry, event organizers continue to insist that the festivities are a private island affair and, as we discovered at the airport, they have gone out of their way to shun the world's spotlight. The result has been a cultural tug-of-war between the islanders who wish to embrace and encourage tourism and the old-school 'Vikings' who do not. As a soothmoother to the tips of my toes, I was about to find myself in the thick of it all.

It was the morning of the festival, the sun had made an unseasonable appearance and I had agreed to give BBC Radio Shetland a short interview. I made my way to the little BBC office and introduced myself to the presenter, who ushered me into the tiny studio.

Before going on air, I asked him what he thought of the festival.

'Well,' he said, before pausing for a moment. 'I'm not sexist, bigoted, racist, homophobic or xenophobic, so it doesn't really appeal to me,' he continued, with a smile. And then before I could say anything in reply the radio jingle had been played and we were live on local radio. 'Welcome to BBC Shetland,' he announced. 'Today I'd like to introduce our guest, Ben Fogle.'

I enthused about the islands and the people and told of my excitement at the prospect of witnessing Up Helly Aa.

'I hear you'll be joining a squad in the procession,' the presenter dropped in, his eyes twinkling with barely concealed glee. 'Joining in?'

I could hear the scoffing and choking as the Vikings digested this nugget of unpalatable news. It was bad enough that outsiders watched the festivities, but no soothmoother was ever, and I repeat ever, to even entertain the idea of taking part. Here we go, I thought: I'd been well and truly stitched up on this one. I skirted around the subject like a government minister. 'Er, um, well, nothing's firmly planned. I'm just going with the flow, I'll see what happens, ha, ha, ha, er, um, you know . . .'

The interview hadn't even finished before the telephone started ringing. It was the committee: I'd been banned from Up Helly Aa. 'Sorry, mate,' lamented the presenter. It was too late. The damage had been done. The Vikings had completely grounded me. I left the studio and moped along the streets, bewailing my predicament and cursing the Vikings. (What have the Vikings ever done for us, anyway?) I could still watch from the sidelines, I supposed.

All day long the beer flowed as the party got under way, and as dusk fell, the excitement reached fever pitch. There was a truly carnival atmosphere as thousands crowded the narrow streets on their way to find the best vantage point for the fiesta, bundled against the cold and clutching warming bottles of whisky. Resplendent in their horned helmets, axes and shields, the Guizer squad assembled near their galley. The grumpy Guizer Jarl had been given the freedom of the town for the day, during which it was his responsibility to lead his squad and nearly a thousand Guizer mates on a marathon twenty-four-hour pub crawl.

'From grand old Viking centuries, Up Helly Aa has come . . .' sang the Guizers as they prepared. As night drew in, the Jarl's army of heavily costumed men formed ranks in the darkened streets, each squad dressed in their own colours and costumes. There were squads of identical schoolgirls, pink dogs, Tony Blairs, Superheroes, Eminems, each one with its own six-foot torch, topped with paraffin-soaked sacking. Elaborate disguise was part of the island's prehistoric fertility rites. Medieval Shetland Guizers were called skeklers and wore costumes of straw. I doubt the founding organizers ever envisaged the Guizers all dressed as Michael Jackson with codpieces, but that's progress for you, I suppose.

On the stroke of 7.30 p.m., a signal rocket – in this case an emergency shipping flare – burst from the town hall. Street lighting was extinguished and the crowd hushed as the torches were lit from an enormous fire pit and the amazing blazing procession began.

The flicker of torch flames bathed the streets in an eerie orange glow, casting strange shadows across the rooftops, a thin plume of smoke giving the overall atmosphere of a town under siege, as a thousand Guizers began their long march snaking their way through the silent town. It made for an achingly surreal sight as grown men with their heavy Viking features and generous frames marched through the streets with their blazing torches, dressed in pigtails, miniskirts and suspenders. A band marked the pace as a river of Bart Simpsons and Britney Spears look-alikes marched forward a mile astern of the Guizer Jarl, standing proudly at the helm of his replica longship.

For an hour, the Jarl's burly squad of Vikings heaved and hauled the ship through the streets, shiny helmets reflecting the orange glow of their torches. It was primeval and beautiful at the same time. Eventually the galley was settled on the ground where it would be sacrificed – in this case, a school playground. The Guizers circled the dragon ship in a slow-motion Catherine wheel of fire. Thousands of islanders hung from windows and crowded the roofs to catch a glimpse of the entrancing sight. The Jarl then raised his axe and shield high into the air and a bugle sounded the call for the torches. A rain of fire descended on the wooden galley as the army of marauding schoolgirls and Bart Simpsons hurled their torches high into the air. It was an incredible sight as the forty-foot ship that had taken a year to build burst into flames.

The blaze soared above the ship's mast, casting orange fiery shadows across the mesmerized faces of the spectators. As the fire destroyed the longship, the crowd sang 'The Norseman's Home', a stirring requiem that brings tears to the eyes of the hardiest Viking. It was an astonishing spectacle.

Tears of wonder became tears of mirth as the night rolled on and the marauding Vikings marched on to a dozen different halls in rotation. At each hall, the forty different squads were each to perform an 'act' or 'show', a choreographed performance usually reflecting local events or politics – lampooning local news is integral to Up Helly Aa. I had been reliably informed that this year's most popular theme seemed to incorporate a local dignitary who had lost his licence after running over a dog while driving under the influence. Every Guizer had a duty to dance with at least one woman in each hall before taking another dram, chased down with some mutton soup diligently made by the otherwise ostracized womenfolk. Up Helly Aa is a very male affair.

I had been specifically barred from the halls, but I had a ticket and I decided to take my chance. After all, there were a dozen halls, a thousand Vikings and an even greater supporting cast of spectators. A long line snaked through the playground and down the street as Shetlanders queued up for the first hall on the circuit. As I reached the door, I noticed a large notice, which read: 'NO ENTRY: BEN FOGLE, VISITING JOURNALIST'.

What next? A barrel of flaming tar left outside my bedroom door? I wondered whether I was in fact in an imperial outpost of Zimbabwe. It all seemed a little extreme, to say the least. What had happened to liberty and free speech, let alone old-fashioned hospitality (have you ever noticed how close that word can be to hostility?)? Maybe it was all a front for Shetland's secret weapons of mass destruction programme. But there it was, written in plain bold letters. I spent the rest of the evening sitting on a wall outside the hall, listening to the frivolities. It was, to say the least, a slightly anti-climax end to the world's greatest fire festival, and a slightly sobering one, despite the company of my small bottle of whisky. I didn't begrudge the islanders – after all, it was nothing to do with them – but I felt ostracized and humiliated as I left the town and headed back to my pencil bed.

Lerwick was a ghost town the following day as I drove through its virtually deserted streets. I passed a handful of Vikings who had passed out on the pavement and I spotted a lone man still wearing his Tony Blair mask and staggering along the road. I was on my way to explore some of Shetland's celebrated scenery.

177

Our plane had landed through thick cloud on arrival and I had caught only a brief glimpse of the stunning coastline, its gently sloping hills and the grazing land of its interior.

Mainland is an enormous island of 357 square miles. The sky is vast and the landscape barren yet truly breathtaking. The first things you notice are the enormous skies and the great open spaces of the hinterland, but I soon found myself drawn towards the magnificent coast and the seemingly infinite variety of its forms: towering cliffs, rocky promontories, sea caves, far-reaching sandy bays, beaches that go on for ever, sand dunes, sea marshes and sea lochs that creep their way inland away from the winds that rush over and around the shores. Like so many of the Scottish islands, Shetland is a nature lover's paradise with an abundance of birds, fish, flora and fauna. And if you're into your puffins, you're in luck because there are nearly a quarter of a million of them living here, as well as about 36,000 gannets, together with otters, dolphins and harbour porpoises. On a good day, you may be fortunate enough to spot one of the orca 'killer whales' that make frequent visits to Shetland's waters. The Guizer Jarl had obviously barred the orcas and the otters from making an appearance before me or any of the other soothmoothers who had dared to come and enjoy the beauty of the islands, but I'm sure I saw what was either a porpoise or dolphin a little out to sea.

The island also boasts a staggering 6,000 archaeological sites that stand as memorials to the long and colourful history of Britain's most northerly communities. I would love to return to Shetland one day – on my own terms, that is, and not on those of a man in a cow-horned Viking hat – because you could spend several months here and still not visit every corner of interest, or even every single island. I would be lying, however, if I said I was sad to leave Shetland on this occasion. It had been a depressing experience for me. I came to celebrate the islanders' remarkable culture for a couple of days, only to be shunned, rejected and humiliated by the festival's organizers.

There was a collective sigh of relief from the assorted press as we took off from Sumburgh airport. Most Scottish newspapers had led with the story about Jarl Peter Fraser's xenophobic outburst in front of the world's media. And I soon discovered that my experience had not been unique. Bouncers had been dispatched to keep the 'invited' new agency photographers at bay and the Norwegian film crew had even resorted to making an exposé about their treatment, as they had been given no access to film the event itself.

I could understand the protectiveness the organizers felt towards their heritage and their festival, but why the reluctance to share it and parade it? Were they not proud of it? Of course one man does not necessarily speak for a whole people, and I couldn't help wondering whether the Jarl's sentiments were just his personally held views or representative of a more general narrow-mindedness or, even, genuine xenophobia.

I chuckled as I read a story in the *Scotsman* that

reported how the local tourist office had decided to drop this year's Jarl and his squad of Vikings as ambassadors to promote Shetland as a tourist destination. Shame, I thought, as I imagined the marketing impact on a curious public of a grumpy Viking with his slogan 'Soothmootherfookeroff!'

# Heligoland

'German Bight, four to five gusting six, squally wintry showers, visibility poor,' the voice crackled over my transistor radio. BBC Radio Four's shipping forecast can be rather soothing when you hear it from the comfort of your own home, a little like a cup of Ovaltine before bedtime. Most of us don't really have a clue what it means, but it's somehow reassuring nonetheless. That's not the case, however, when it translates into layman's language as 'Bloody wet, windy and wavy' and you're in the midst of it aboard a ferry, as I was, in the middle of the German Bight in the North Sea.

The German Bight is a wild 20,000-square-mile area of sea and coast that stretches between two headlands close to the Dutch island of Texel at the southern end and up to the Jutland port of Esjberg in Denmark in the north. German Bight has always struck me as a curious name. It was only fifty years ago that it came into being. In 1955 government meteorologists representing countries bordering the North Sea convened to discuss the renaming of the region's sea areas. Dogger was halved and its north-eastern section named Fisher, while Forties was also divided and its northern half named Viking. At the meeting the

German delegation demanded that the Heligoland Bight should be renamed the German Bight. To understand the reason behind this demand you have to know the extraordinary history of a Germanic island with very British roots: Heligoland, the island that the British Empire abandoned and then tried its best to forget.

This curiosity of an island lies some thirty miles from the coast of Schleswig-Holstein and Lower Saxony on Germany's North Sea coast. It rises like a red-gloved fist from the swirling waters of the North Sea and its cliffs tower 200 feet above the sea, their ochre sandstone vivid against the drab grey of the ocean. The sedimentary rock that gives the island its distinct physical character is one of the many odd things about Heligoland: you cannot find a similar geological formation anywhere else along the North Sea coast. The island used to be far larger, with dunes and low-lying marshes stretching virtually to the European mainland, but rising sea levels and continuous wave attack have reduced it to a little red rump of just 500 acres.

Legend has it that St Willibrod, an English missionary from Lindisfarne, was shipwrecked in Heligoland in 699. There are few records but it seems that the name, which means holy island, dates from this period. Soon afterwards, a number of Norse chieftains claimed this storm-battered island as their own. The island finally assumed English sovereignty when in 1016 King Cnut of Denmark also became King of England, bringing Heligoland with him as a sort of royal dowry.

The island remained nominally under Danish rule, but Heligoland's farming and fishing community were left largely to their own devices, and to run their island as they always had.

When you look at it on a map of Europe and see its position in the corner of the North Sea near the mouth of the River Elbe, it is not difficult to understand how it came to have great military importance, particularly in the two great conflicts of the twentieth century. It was not until the Napoleonic Wars, though, that Britain first began to appreciate Heligoland's strategic importance. It was seized by the British Navy from the Danish in 1807 and formally ceded in 1814, when it became Britain's smallest imperial outpost as well as her only colonial possession in northern Europe.

During the Napoleonic Wars Heligoland played a crucial role as the forward base for the officially endorsed smuggling of contraband to the Continent and as a centre for intelligence gathering. After the Navy had seen off the Danish occupiers, Britain moved a sizeable garrison on to the island, but in the years that followed the British developed a distinctly laissez-faire attitude to it. For Britain, a major world power with more island colonies scattered across the globe than it knew what to do with, there was nothing particularly special about Heligoland. In Victorian times its people were seldom invited to colonial gatherings. Advertisements in London newspapers boasted that you could reach the island in just thirty hours from London, but still very few people visited. For neighbouring

Germany, meanwhile, the island was both attractive and tantalizingly close to its shores. Heligoland slowly became more German than English. It became a popular health and bathing resort for the Germans and received a number of distinguished visitors over the years, including the composers Mahler, Liszt and Bruckner, the latter being inspired to write an orchestral piece about the island. Kafka, the author of *The Trial*, also stayed on the island while Heinrich Hoffmann sat down and wrote the words for the German national anthem 'Deutschland Uber Alles' one evening after a long walk around it in 1841.

Some years before he was crowned, Kaiser Wilhelm II visited Heligoland and vowed to make it German. Bismarck, his chancellor, coveted the island for its strategic value for the proposed Kiel Canal linking the North Sea with the Baltic. Indeed, he once suggested to the British Prime Minister William Gladstone that the island might be exchanged for an enclave in India called Pondicherry. The offer was refused. In the nineteenth century the preposterously untrue German claims that Heligoland had originally been German went ignored and unchallenged, such was Britain's indifference to the little outpost. Britain was far more interested in Germany's territories in East Africa, notably Zanzibar and Uganda, which she hoped one day to acquire for herself.

Although there is a small landing strip on the island, the vast majority of people reach Heligoland (known locally as Helgoland) by ferry from Hamburg, Germany's principal port on the River Elbe. During the summer months the boat service is fairly regular, but for the rest of the year the island remains largely cut off from the rest of the world. It was early spring and I was on the first boat of the season to an island that barely anyone in Britain has heard of today, let alone visited. The River Elbe is one of Europe's major waterways leading into central Europe. Once creating an artificial boundary between West and East Germany, this 1,000-mile-long river is linked to canals that reach as far inland as Prague. The Elbe–Lubeck and Kiel canals link the Elbe to the Baltic, while the main river flows into the North Sea at Cuxhaven. Today the river is a vital shipping route, along which pass the majority of Germany's imports and exports, via Hamburg. Largely destroyed in the Second World War, today Hamburg is a bustling city in the throes of regeneration. Having undergone an astonishing facelift in the last ten years, it is crammed with *uber* bars, designer hotels and contemporary architecture.

A weak sun was rising as the ferry pulled away from the bustling quayside. We sailed past heavy industrial sites, dry docks and oil refineries. While vast cargo ships and car carriers squeezed by, towering overhead like ugly tower blocks, and an assortment of ferries, fishing boats and pilot ships buzzed up and down the busy waterway. The estuary was veiled in a thin pall of early-morning mist, which streaked across the murky water like cotton wool. Heligoland is one of the last bastions of duty-free shopping in Europe and each year tens of thousands of German day-trippers head there to stock up on cigarettes and alcohol – on the equivalent of the English 'booze cruise' to France. Today out of a hundred or so passengers a dozen hardy ones, wearing what seemed to be a regulation uniform of tight jeans and leather jackets, huddled on the tiny viewing deck, chain-smoking and sipping espressos. The bitter cold chilled the bones.

For several hours we sailed downstream in a fug of cigarette smoke as excruciating German pop music grated the ears from the ferry's tannoy. We cruised past dozens of the small settlements and factories that clutter the riverbanks before eventually reaching Cuxhaven, where the river meets the great North Sea. Cuxhaven is one of the largest fishing ports of Germany, with a population of over 55,000, most of whom are employed directly or indirectly in the fish-processing industry. The town is also a North Sea beach resort, a health spa and a base for sailing expeditions to the Friesian Islands.

Cuxhaven disappeared over the horizon and for several hours we beat into brisk North Sea wind. The unseasonable sun, now high in the blue sky, gently warmed my skin. The ruby-red face of Heligoland soon appeared in the distance, her sheer cliffs on one side of the island dominating the horizon. We were still ten miles away, but I could make out the enormous mast and tower that stand on top of the island like a steeple on a sandstone cathedral. From a distance she looked like a watery version of Australia's Ayers Rock. It seemed incredible that this bizarre island, more than three hundred miles from England, had once been a British possession, the Gibraltar of the North Sea.

The British ruled Heligoland without incident until 1890, when they used this increasingly strategic island to barter with the increasingly imperialistic Kaiser Wilhelm II. Desperate to prevent further Teutonic (German) expansion into East Africa, Lord Salisbury, the Prime Minister, agreed to cede Heligoland to the Germans in return for strategic lands in Africa, including Zanzibar. The Heligolanders were never consulted as their cultural identity and nationality were swapped in an international battle of imperialistic ambition. Germany had agreed to give up her East African aspirations in return for a seemingly useless isle. However, with Germanic efficiency Heligoland was soon transformed to make the most of its strategic and geographical position: a naval base, a large harbour and dockyard, and coastal batteries were all constructed, turning Heligoland into a fortified island that would soon threaten the very heart of Britain.

In Britain, this audacious and quite unprecedented territorial swap provoked protest. Even Queen Victoria furiously remonstrated that the 2,000 inhabitants of this sophisticated island were being callously sacrificed. In a telegram to Lord Salisbury in 1890, she wrote: 'The people have always been very loyal and it is a shame to hand them over to an unscrupulous despotic Government like the German's without first consulting them. It is a very bad precedent. The next thing will be to propose to give up Gibraltar: and soon nothing will be secure.' It still seems amazing today that the Heligolanders only found out about their fate after the event, via the newspapers.

On 8 August 1890 this unique deal was formalized in a simple ceremony at which the British and German flags were lowered simultaneously in East Africa and Heligoland. There was no pomp – this was certainly no Hong Kong handover. The gesture would have profound ramifications for the North Sea isle, which soon found itself in the middle of the first skirmish of the First World War. The First Sea Lord, Admiral Fisher, declared that Heligoland had become 'a dagger pointed at England's heart', as the once British island was turned against mainland Britain and the Battle of Heligoland Bight raged in the island waters. Heligoland also became embroiled in the first organized seaplane attack of the war with the Royal Naval Flying Corps's participation in the Cuxhaven Raid. The islanders were interned in German prison camps, but they never gave up the hope that they would one day, once again, be the loyal subjects of Queen Victoria.

In the Great War the island was in constant use by German naval forces, but all the military works and installations were demolished shortly afterwards in accordance with the Treaty of Versailles. For a few years the island became a popular tourist resort until the launch of Project Hummerschere, an ambitious scheme to construct a German equivalent of Scapa

Flow in the Orkneys, Britain's chief naval base in the two wars. So important was the project that in 1938 Hitler himself visited the island and it was perhaps unsurprising that the island became the focus for Britain's first mass night bombing raid. In fact, the very first bomb to land on German soil during the new conflict landed on Heligoland. The islanders remained on the main island throughout the six years of the war, but on 18 April 1945 over a thousand Allied bombers attacked the island, leaving nothing standing. The civilian population was protected by rock shelters, and most of the 128 people who died that day were members of anti-aircraft crews. Today you can still see the craters and distortions to the high plateau caused by the bombing. The island was evacuated the following night. At the end of the hostilities, Britain 'confiscated' the island it had handed over to Germany fifty-five years earlier.

It went largely unreported or unnoticed at the time, that the Heligolanders who had returned to their homeland were exiled to mainland Germany to allow the British to use the island as a bombing range for high-explosive and chemical weapons, and even as a test site for various elements of Britain's prototype atomic bomb, which it did until 1952.

As our ferry pulled into the small sheltered harbour, past the island's dedicated lifeboat and dozens of small wooden fishing boats, remnants of Project Hummer-schere and parts of the destroyed U-boat shelter were visible in the shallow water. The ferry disgorged its cargo of leather-jacketed tourists; and a handful of islanders, wrapped up against the biting North Sea wind, watched our arrival. Next to the jetty was a *Musik-pavillion*, one of the island's ubiquitous glass structures in which visitors can shelter from the wind. Rows of brightly coloured homes lined the waterfront, giving the island a slightly Nordic feel.

As on Sark, there are no cars on Heligoland, and not even any bicycles. A few electric vehicles like milk floats are used to haul supplies around the island along its single-track roads. Heligoland has changed little since Victorian times, even though the town, the harbour and the bathing resort of Dune were completely rebuilt only a few decades ago so that the island could be reinhabited. The high street, where once stood warehouses crammed with merchandise for smuggling to the mainland in defiance of Napoleon's continental system, is now brimming with duty-free shops piled high with alcohol, cigarettes and scents. On the outside walls of the hotels and guesthouses hang huge maritime pictures – some even of old British merchant ships. Traditionally, despite Heligoland's constitutional links to Denmark, Britain and Germany, the islanders regard themselves and their island as a distinct people and nation.

The island is roughly divided into the Oberland and the Unterland, or the higher and lower parts. The lower part is largely given over to the tourist industry and the majority of the island's 2,000 residents live on the protected side of the plateau. Nearly all the houses have small gardens, colourful window boxes and hanging baskets. There is a street called Gouverneur Maxse-Strasse, named after an ornithologist who was part of the island's British administration during Victorian times. Near by is the rebuilt St Nicolai Church, from the ceiling of which hangs a model ship, a replica of the one donated by Governor Maxse. On the wall outside there is a metal plaque dedicated to Queen Victoria.

Shifting economic circumstances have had their effect on Heligoland: in particular, cheap, low-cost airlines to other European destinations have hampered tourism. Fishing is still the most important industry on the island, with lobster the speciality. Some of the bombs from the 1945 raid narrowly missed the lobster ground on the weather side of the island and are still there, trapped in the fissures of the rocks. Every now and then lobstermen accidentally haul one to the surface and the German Navy makes it safe. But these explosives were mere fireworks compared to what Britain had in store for this stormy atoll. On 19 July 1946 a team of Royal Marine demolition experts arrived with a consignment of 455,591 boxes of TNT. 'The greatest non-atomic explosion in history' was how *The Times* described the 'big bang' that took place on 18 April the following year. The explosion was so powerful that it registered on over a hundred seismographs across Europe. Operation Big Bang nearly wiped the island off the face of the world. Indeed the plan was to destroy the island once and for all, but somehow it survived the 6,800 tons of explosive that shook the North Sea that day. I walked along the road that runs through the vast crater left by the 1947 big bang. In the middle there now stands a hospital with an enormous red cross painted on its roof, with a clinic for sufferers of Parkinson's disease.

It wasn't until 1952 that Heligoland was returned to Germany and the Heligolanders were finally allowed to return to their scarred and cratered island. Few buildings remained; it must have been heartbreaking for the islanders when they returned to find their island in ruins and almost unrecognizable. But the islanders picked up their lives again, and with grim determination and stubbornness they once again started to eke out an existence from the little that was left.

Heligoland is a beguiling place, its nature determined predominantly by the rhythms of the seasons, and more specifically by the ocean. It is perhaps not surprising that there is a local saying '*Nordsee ist mordsee*' – the North Sea is murderous. From the top of the lighthouse I could make out the faint outline of the north German coast. A more fascinating sight was the island's plateau, rutted and cratered like a grotesque golf course.

I wandered along the coastal path that runs the circumference of the island. On the north side, where the cliffs tower above the lower town, reinforced-steel cages had been erected around the houses to protect them from falling rocks and landslides, caging their gardens as if they were prison yards. Away from the town, the precipitous cliffs fall to the ocean several hundred feet below. The North Sea has taken its toll on this island over the years, biting into its red cliffs and eating away at its very foundations. So profound was the coastal erosion that authorities were forced to intervene and build a wall around the entire island. This feat of engineering gives the island the overall impression of a sea fortress, a castle in the ocean surrounded by a dry moat.

Along the coastal path, the various disused gun mountings have been made into bird watching platforms. The sheltered side of the island is home to Heligoland's allotments, small parcels of land owned and farmed by each family. The island is blessed with above-average sunshine, and the islanders grow an astonishing array of fruit and vegetables on these high cliffs. Each plot has its own small workshop, built of driftwood. As I continued my blustery walk I was struck by the number of island flags I saw, flapping from crude masts in the fresh breeze. The red, white and green of the flag is defined by a Heligoland saying, 'Green is the land, red is the strand, white is the sand.' Crime, it would seem, is a rare occurrence. A note in German on the island noticeboard from a 'Roger Russel' asked the person who 'borrowed' his spade to kindly return it.

Evidence of the island's history is everywhere, not just in the pot-hole-riddled landscape but in smaller details such as the signpost with one arrow pointing towards England with the distance given as 55 kilometres, and the other to Zanzibar with the distance 7,300 kilometres. Heligolanders will not easily forget the day Britain betrayed them. On the evening of 9 August each year many of the 2,000 islanders gather to watch an annual pageant commemorating Governor Barkly's handover of the territory to Germany. It is not a celebration but a memorial.

By mid-afternoon, a procession of German day-trippers was streaming towards the little port, laden with scent and bottles of whisky, some even pushing island shopping trolleys brimming with their bounty of tax-free goods. I watched as the ferry pulled from the harbour, once again enveloped in a fug of tax-free cigarette smoke. I was sure that I could hear the island breathe a sigh of relief as it was once again left to itself. Suddenly I felt as if I was the only person on the island as I walked along the deserted streets towards my hotel, the island's newest residence, the Atoll, a big, brash over-designed hotel rather out of keeping with the island's low-key atmosphere and relaxed pace of life. A huge glass atrium dominated the entrance foyer, with bright plastic spyholes protruding from the floor like barnacles, offering a voyeuristic peep into the swimming pool below. A glass lift shuttled between the hotel's six floors, its windows streaked with salt. Surprisingly for a six-storey hotel on an island's edge, my room had no sea view; instead I had a delightful vista of a row of duty-free kiosks. The room itself was very bright, very yellow and very plastic, a bit 'porn star' with its glass-encased shower in the middle of the room, a little sixties with its moulded plastic sofa and a little 'Ian Schrager' with its colour-coordinated floors. It was more Legoland than Heligoland.

That evening, in the interests of research of course, I found myself ensconced in one of the island's small pubs with a group of local fishermen. Notwithstanding the language barrier, I was struck by the tough and independent-minded population: a close-knit community of seafaring folk, strong, stoic and proud. I asked if they felt English or German.

'We are neither,' announced a moustachioed lobsterman in grubby overalls and a thick seafaring jumper. 'We are Helgolanders.'

'But who do you feel more loyalty towards?' I pressed.

'Britain,' he said, giving me a wink before raising a glass and announcing, 'To Helgoland.' I wondered whether he was saying it for effect; I got the impression that for some of the islanders Britain's betrayal of their grandparents was still raw.

The next morning I decided to explore nearby Sandy Island. Half a mile from Heligoland, this little strip of land is a nature reserve that doubles as the island's occasional airport, home to the world's smallest 'national' carrier, Heligoland Airways. The island is defined by vast white sandy beaches and dunes, which make it a popular sunbathing spot all year around as well as a haven for hundreds of seals that cluttered the beach when I arrived, soaking up the weak rays of sunshine.

In addition to the dozens of snoring, grunting seals, there were a dozen or so Germans, sheltered from the wind by tiny deck-chair cubicles. These brightly coloured 'huts' with their hoods and swivelling bases could be turned and adapted to provide minimum wind and maximum sun in a very practical way that only the Germans could have come up with. I sat in one of the 'sun boxes' with a bottle of German beer and immersed myself in Erskine Childers' novel *The Riddle of the Sands*, which is said to be based on the island.

You cannot help but admire the unwavering obstinacy of Heligoland and its people, I thought. Everything about the island speaks of stubbornness, a refusal to yield to outside forces: the sea defences built to protect her from the ravages of the North Sea, the cages erected behind the terraced houses in the lower town to protect them from rock falls, the duty-free shops stocked full of luxury goods to attract and entice the euros so crucial to the island's future. Heligoland had survived betrayal by Britain, two world wars, evacuation, ferocious bombing raids, the 'big bang', economic downturn and thousands of years of winter storms, and somehow she was still here, clinging on against all the odds.

# Sealand

I spent the entire journey from London to the Essex port of Harwich practising my royal greeting etiquette. 'How do you do, Your Royal Highness . . . It is a great pleasure to make your acquaintance, Your Majesty . . . Good day, Prince, sir, Your Highness . . .' Was I to bow? Should I wait until I was spoken to? Could I touch or even have eye contact? I had never met any royals and had no idea how to behave in their presence. Within a few minutes I would be shaking hands with a member of the world's most unusual monarchy, Prince Michael of Sealand, Prince Regent and de facto ruler of a weird little kingdom in the North Sea measuring just 546 square yards.

'You must be Ben,' beamed the Prince in a gruff Essex accent as he stepped from the passenger door of a convertible BMW, and squeezed my hand to within an inch of its life. He sounded more like Mike Read than Prince Charles. I quickly axed the curtsy and bowing plans and plumped for a common-or-garden 'Hello'. A young man, no more than about sixteen, appeared from the driver's seat. 'Meet my sons, Prince Liam and Prince James. James is my driver until I get my licence back. The rozzers clocked me at 140 miles per hour,' Prince Michael explained. I shook the young

197

Prince's hand; I could see the other in the passenger seat.

So not much pomp or ceremony in Harwich, then: no crowns, no red carpets, no footmen, no powdered wigs. This certainly wasn't the regal welcome I had anticipated, but then Prince Michael isn't exactly your typical royal and Sealand is no ordinary sovereign territory. The principality of Sealand started life as Roughs Towers, one of four Maunsell seaforts built for the Navy around the Thames estuary in the Second World War to act as anti-aircraft batteries and anti-mining bases (another three Maunsell forts were built for the Army). The forts were constructed in dockyards and then floated out to sea, flooded and sunk into their locations. A gun platform sat atop two giant hollow legs that provided accommodation and storage facilities for the soldiers stationed there. The forts prevented the Germans from laying hundreds of mines in the strategically vital Thames estuary and between them shot down twenty-two aircraft and thirty doodlebugs.

The forts were maintained during the Cold War in anticipation of another mighty sea battle, but as the threat of a Third World War diminished and relations with the West warmed, the towers were abandoned to the seagulls. However, by the middle of the 1960s a number of pirate radio stations had began broadcasting from offshore ships around UK waters in order to skirt Britain's strict monopolistic broadcasting regulations. Radio Caroline started a rush for offshore bases, inspiring the likes of Screaming Lord Sutch to begin rogue transmissions to the swinging masses of the sixties. It wasn't long before the Maunsell forts became over-sized turntables.

One of those quasi-legal stations was Radio Essex, the first twenty-four-hour local radio station, which was started on Fort Knock John by one Roy Bates, an ex-army major who was then running an inshore fishing fleet. Roy had fought in the Spanish Civil War and in the Middle East, and had been wounded in action fighting in Sicily. After falling foul of the authorities for his illegal broadcasts, Roy became discontented with authority and, always looking for a challenge, he decided to start his own country. The idea of Sealand came to him after the government amended the definition of what constituted international waters. Arguably Sealand is the only country on the planet to have been born over a few pints in the local pub. Bates knew that there were other platforms in the North Sea, and so he began consulting law books and international lawyers and was delighted to discover that there were no obvious legal obstacles to his daring project. After a brief disagreement with Radio Caroline, which was using the fort as a replenishment platform for its ship, Bates proclaimed the island his own state, moved his family on to the fort and set about renovating his new home and kingdom. On 2 September 1967 he declared the fort independent, and claiming the ancient law of *jus gentium*, he bestowed upon himself and his wife the titles Their Royal Highnesses Prince Roy and Princess Joan of Sealand. After raising the flag of the principality of Sealand for the first time, he set about printing passports (of which there are 300 authentic ones in circulation today), as well as Sealand dollars, coins and stamps, all embossed with the profiles of the world's newest Prince and Princess. Recognizing the commercial possibilities of the young nation, he looked into creating a tax haven, complete with hotel and casino.

The British government, then under Harold Macmillan, took a dim view of this new sovereign territory a few miles offshore. However, when a naval cruiser sailed close to Sealand and was promptly warned off by shots fired from the platform, the incident landed the newly declared Prince in trouble with the law once again; but at the subsequent court case Mr Justice Chapman ruled that as Sealand was outside the three-mile limit of British territorial waters, the United Kingdom could not claim sovereignty over the fort and he and the court had no jurisdiction over it. His ruling tacitly acknowledged that the royal family Bates was perfectly entitled to their sovereignty.

I had waited for nearly a year for permission to visit the world's smallest country. From the outset, prospects hadn't looked good. An official notice on its website warned, 'Due to the current international situation and other factors, visits to the Principality are not permitted ... Accordingly, the application list for visas is for the time being closed.' The notice continued with a glimmer of hope, though, adding, 'All requests are carefully considered and a delay of at least thirty days may be expected before a decision is made.'

I submitted my application, more in hope than in expectation, explaining about my journey to various islands around the UK's coastline. Thirty days became three months, and then seven months, and I had all but wiped Sealand off my proposed itinerary when an e-mail dropped into my inbox notifying me that my application had been received and was being considered. It was curiously reassuring to know that the cogs of bureaucracy worked just as slowly in a nation with a population of five as they did elsewhere in the world. A month later I was surprised to receive a notice that my application had been accepted and my visit could go ahead – but then a few weeks later it was cancelled. And so it continued for a further six months: each attempt to visit was scuppered at the last minute. 'The Prince will be away on business ... The weather forecast is for gales ... The Prince is sailing in Scotland ...' That's the problem with royals – always busy.

I had almost forgotten about Sealand when, finally, I received a genuinely positive and detailed e-mail informing me that I had been 'cleared for a trip to the Principality'. The trip was to take place in just a few days' time, on the following Monday. I was told to meet 'Immigration' at the Pier Hotel in Harwich at 0900 promptly, with my passport and visa forms. At last, it seemed, I was going to Sealand.

\*

In 1978 Sealand was subjected to what Prince Roy called 'a terrorist attack', an event that shook the sovereignty to its foundations and led to newspaper headlines around the world. Prince Roy and Princess Joan were away on business when Sealand was invaded by a group of men from Holland and Germany. They thought the old fort was unoccupied, but when they arrived, they discovered that the Bates's young son Michael had been left in charge. He was taken hostage and incarcerated while the occupiers brought in reinforcements by helicopter from Hamburg. The young Prince was eventually deported to the Dutch coast while an assortment of ne'er-do-wells settled into their newly captured territory.

The Bates family were soon reunited and after enlisting armed assistance they set out to liberate the principality and reclaim their kingdom in a daring, Bond-style helicopter raid. Sweeping low across the waters, they caught the occupiers by surprise and quickly recaptured the fort. While most of the raiders were deported to the UK, where they were promptly arrested, one, a German lawyer named Gernot Putz, was charged with treason. Putz owned a Sealand passport, given to him for his services to the country. A court was convened, and the waiting press, including reporters from such august publications as the *Financial Times* and the *New York Times*, watched as Putz was sentenced to twenty years in prison and fined £20,000. The German authorities complained to their British counterparts and asked them to intervene, but as

Britain did not recognize Sealand there was little they could do. Germany eventually sent a diplomat by helicopter to negotiate with Sealand. In the end, after forty-nine days in captivity, during which he was forced to 'wash the loos and make the coffee', Putz the pirate was given a royal pardon and released.

The next couple of decades passed relatively un-eventfully for Sealand, except when East London's notorious Kray brothers attempted to seize the fort in a rigid inflatable, which was promptly sunk with the help of a swiftly concocted Molotov cocktail dropped by a young Prince Michael. Happy in the knowledge that peace was now abroad in their little kingdom, Prince Roy and Princess Joan retired to Spain, leaving their son Michael in charge of Sealand's affairs.

'Were you scared when you were taken hostage or when you re-took Sealand that day?' I asked Prince Michael during our meeting.

'I was more infuriated,' he replied. 'In fact I was furious. By the time I got back to my father we'd learnt that they were planning to put ten Belgian paratroopers on here the next day, so we had to move pretty damn fast. My father phoned a friend of his who was a stunt pilot for James Bond films and we surprised them by flying low over the sea and coming up suddenly under-neath. We all stood on the skids of the chopper with guns and coiled ropes,' he said, a smile widening across his face. 'It was very close to becoming a firefight, but eventually we retook the fort without a shot fired.'

\*

'You'd better clear customs and immigration,' HRH told me, giving me a wink and pointing me to the hotel's deserted bar. In the corner sat a man with a heavily lined face, partly obscured in a fug of cigarette smoke. He wore a sweatshirt embossed with the words 'SEALAND SECURITY'. Blimey, I thought, perhaps this is not just an elaborate game after all.

'Hello,' I said with a smile. 'I'd like to go to Sealand.'

'Passport,' he replied in a gravelly voice.

I promptly handed it over.

'Visa forms.'

I passed them across the table.

'That will be a hundred Sealand dollars.'

Gulp. Where was I supposed to get Sealand currency?

'Is there a bureau de change?' I enquired.

'We accept euros and sterling,' he announced curtly.

I handed over the equivalent of £100 and scurried back to HRH and the young Princes, who were sipping cappuccinos at the bar. I later learned that the gruff immigration official was Lew Schnurr, Chief of the Bureau of Internal Affairs of the Principality of Sealand.

We headed down to the docks. 'I'm afraid Sealand One is in for repairs,' lamented Prince Michael, referring to the fast black rigid inflatable boat on which the Bateses normally commute to Sealand. Our royal yacht turned out to be a very unregal model of transport – a local fishing boat requisitioned by HRH.

*

In 1997 the principality of Sealand was hijacked on the Internet by an alleged international fraud ring with interests in weapons trading, drug smuggling and money laundering. The rival 'Sealanders', based in Germany and Spain and rumoured to be associates of the bungling hijackers looking to get their own back, sold passports, designed uniforms for a phantom army and printed labels for a Sealand whisky before police raided the 'Sealand embassy' in Madrid and detained the Spanish 'regent'. His self-styled government and diplomatic corps were investigated by Interpol and as a result, Sealand unwittingly found itself at the forefront of world news once again.

The first sign of a Sealand clone came after the 1997 murder of fashion designer Gianni Versace and the suicide of his killer, Andrew Cunanan, on a Miami houseboat. Police discovered that the German businessman who owned the boat had a 'Sealand' passport and even diplomatic licence plates. Bates was puzzled because his principality had issued only a few hundred passports to friends and collaborators and didn't know the German businessman in question.

Suddenly thousands of black-bound 'Sealand' passports embossed with the Bates family coat of arms – two crowned sea creatures – began turning up across the world. Nearly four thousand were sold in Hong Kong alone for £1,000 apiece as locals scrambled to obtain foreign documents before Britain handed over the colony to China in 1997. A 'Sealand' website appeared which boasted of 160,000 citizens and listed a number of foreign embassies. Sealand didn't have citizens or even a website at the time. When the Prince and Princess contacted the website, someone claiming to have close ties with Spain's King Juan Carlos wrote back insisting he was 'working in the interest of my family and doing all kinds of wonderful things'.

The mystery soon began to unravel when Spain's paramilitary police arrested a flamenco nightclub owner for selling diluted gasoline at his Madrid filling station. Identifying himself as 'Sealand's consul', he produced a diplomatic passport and tried to claim immunity from prosecution. Police raided three 'Sealand' offices and a shop that made 'Sealand' licence plates. These shady operatives, police claimed, had sold diplomatic credentials to Moroccan hash smugglers, negotiated with Russian arms dealers and tried to channel millions of dollars through three Spanish banks for mysterious Russian and Iraqi clients. They offered shipping documents and then university degrees, with astonishing sums up to the value of £40,000 trading hands. About eighty people were accused of committing fraud, falsifying documents and usurping public functions. Police in the United States, Britain, France, Germany, Slovenia, Russia and Romania were all involved.

Of course all this had nothing to do with the Bates family and the real Sealand kingdom. 'They're stealing our name, and they're stealing from other people,' said Joan Bates at the time. 'How disgusting can you get?'

One of the accused, Friedbert Ley, a German businessman who had registered the site in 1998, insisted that he was not involved with the ill-fated invasion but admitted that he had been fascinated by Sealand for a quarter of a century. Germany does not outlaw the sale of 'fantasy documents', but crimes such as the Versace killing and the collapse of East European pyramid schemes finally raised serious questions about Sealand citizenship – for instance, an Austrian couple using a Sealand passport had opened a bank account in Slovenia under false names for laundering profits. Spanish investigators discovered that Ley had lowered his profile and moved 'Sealand's' government to Spain under a man called Francisco Trujillo, who had worked for Ley's roof insulation company near Dusseldorf. Ley's website added a Spanish-language version at about the time of his appointment as 'Sealand's prime minister' in a quasi-cabinet of Spaniards, Germans, Russians and Brazilians.

Unlike the Germans, Spanish police viewed 'Sealand' documents as tools of fraud. Their investigation brought to light the shadowy leaders. They discovered that Trujillo had amassed a collection of tailor-made combat uniforms for a 'Sealand army', of which he made himself colonel. Trujillo was described as a 'mythomaniac', and police discovered that nine of the ten members of Trujillo's government had police records for fraud and, more worryingly, possession of explosives.

But all this paled into insignificance when Spanish

officials discovered that the group had attempted to broker a $50 million deal to send fifty tanks, ten MIG-23 fighter jets and other combat aircraft, artillery and armoured vehicles from Russia to Sudan, circumventing a European arms embargo against the African nation, which was accused of terrorism. To give the gang a veneer of respect, Trujillo had set up a cultural foundation named after the Spanish painter Francisco Goya and attracted some unsuspecting VIPs, including a former chief of Spain's royal household. Embarrassed by the scandal, Spanish officials tried to dismiss the case on the grounds that a fake country's documents can harm only unsuspecting buyers, not Spain's national interest.

Trujillo's 'Sealand' was also trying to buy 1,600 cars and two cargo aircraft on credit. But the biggest deception of all perhaps was the group's bid for 'foreign investment' in a utopian project that entailed expanding the platform for a luxury hotel and casino, business centre, sports complex, medical centre, tuition-free university and – my personal favourite – a Roman Catholic cathedral. Meanwhile, as police pursued the Spanish interlopers, Bates struck the deal he'd been waiting for. Sean Hastings, a 32-year-old American whizzkid of Silicon Valley and Caribbean-based computer ventures, had been looking for a safe offshore headquarters. In a book called *How to Start Your Own Country* he read about Sealand and contacted its founder.

*

After an hour Sealand appeared on the horizon looking like a bizarre sea monster, a sort of aquatic tripod,
a version from *War of the Worlds*. A further half an
hour and I could make out the word 'SEALAND' painted
on the side of the buildings, and the sovereignty's
black, red and white national flag fluttering in the stiff
breeze. We slowed to a halt as we drew near to the
base of one of her stumpy legs. She towered nearly a
hundred feet above us. I scanned her structure for
a ladder or staircase. There was nothing – just the
barnacle-encrusted sheer steel dipping into the stormy
ocean, pounded by waves.

'How do we get up?' I asked, a little puzzled.

'You'll see,' answered young Prince Liam, handing
me a life jacket. (It seemed a tad strange that only now
after we had bucked our way for an hour across a
choppy sea and reached the safety of Sealand did I
need a life jacket.) High above, tiny heads appeared
over the parapet as people manoeuvred what looked
like a very old crane over the side. A thick cable
descended to our bucking boat, on the end of which
was a short plank of wood acting as a sort of bo'sun's
seat.

'Do you want to go first?' asked Prince James. Without waiting for an answer, he guided me into the
contraption. Suddenly I was dangling in the air.
Buffeted by the wind, the boat disappeared below me,
and all I could see was the swelling of an angry sea.
Up, up and up I was hoisted, like a puppet on a string.
Eventually the arm of the winch swung me over the

side of the platform and dropped me on to Sealand. I'd finally made it.

Moments later, I was joined by the young princes and Michael, who were warmly welcomed by their subjects. The Crown Prince was back on his domain. I removed my lifejacket and walked into the building.

'Passport,' demanded a man with heavy wrinkles in a 'SEALAND SECURITY' sweatshirt. It was Lew from the Harwich hotel. How the hell had he got here? I hadn't noticed him on the small boat.

'Have you got a visa?' he asked.

'Of course I have. I thought you just gave me one,' I stammered.

'Enjoy your stay,' he smiled, stamping my passport.

Immigration control, which doubles as the seat of Sealand's government, also happens to be the country's kitchen, from which a huge bay window looks out on to the rusty red deck. A fridge and freezer purred along the wall. Dozens of cans of baked beans lined the shelves, together with bottles of tomato ketchup, boxes of tea bags and cartons of UHT long-life milk. A huge drainage hole in the floor had been blocked with a cork. On particularly stormy days, the cork pops from the hole like a champagne cork and white water erupts through it like champagne foam. Storms are common here and sometimes Sealand is cut off for weeks at a time.

I sat down on one of the plastic chairs and over a cup of tea Prince Michael regaled me with some highly entertaining stories about life and business on Sealand.

Dividing his time between his cockling boat and governing Sealand, he has tried numerous business ventures over the years, ranging from the sublime to the ridiculous. Among many ruses was his attempt to market the legs of the old fort as advertising space to Pepsi and to a marine engine oil company; and he even tried to set up the UK's first pirate television station to be beamed to southern England and presented by Suzan Mizzi, the former 'Page Three' model. None, however, was quite as ambitious as his ventures into the world of cyberspace with a company called HavenCo.

HavenCo promised to revolutionize Sealand – and the World Wide Web in the process – while making millionaires of everyone involved in the project. The story reads like the plot of a Hollywood blockbuster. Ryan Lackey, a shaven-headed 21-year-old MIT dropout and self-taught crypto-geek, became part of a crew of computer adventurers and 'cyberpunks' working to transform the principality into 'a fat-pipe Internet farm' and global networking hub – a pirate Internet site in short. In 2000, with $1 million in seed money provided by a small core of Internet-fattened investors, Lackey and his colleagues, including Sean Hastings, set up Sealand as the world's first truly offshore, almost-anything-goes electronic data haven. The company, HavenCo Ltd, was registered at a London address but headquartered on the 6,000-square-foot former military outpost. Sealand's ageing interior was soon transformed as the huge support cylinders were filled with several million dollars' worth of networking gear: computers, servers, transaction processors, data-storage devices – all cooled with banks of roaring air conditioners and powered by mighty generators.

HavenCo was able to provide its clients with a gigabyte per second of internet bandwidth, at prices far cheaper than those on the over-regulated dry land of Europe. Three speedy connections to HavenCo's affiliated hubs all over the planet – microwave, satellite and underwater fibre-optic links – ensured that the data never stopped flowing. HavenCo promised to provide people with a safe, secure shelter from lawyers, government snoops and assorted busybodies without their getting tangled in flagrant wrongdoing. So if you ran a financial institution that was looking to operate an anonymous and untraceable payment system, HavenCo was for you, but if you wanted to run a scamming operation, launder drugs money or send children's porn anywhere, forget it. Six computer geeks, protected by four security personnel, spent nearly a year working on Sealand, installing the infrastructure. They even planned to fill the machine rooms with a pure nitrogen mix, requiring scuba gear to enter it and work in it, as a measure designed to keep out sneaks, prevent rust and reduce the risk of fire.

Before taking me below to see the HavenCo operations, Prince Michael was keen to show me some of the trappings and memorabilia of state, including one of the rare genuine Sealand passports. I was surprised to see that it had belonged to a doctor from Onslow

Square in London and had been used to travel to the United States, Turkey, Australia – in fact it was filled with stamps. He was then proudly showing me a cache of Sealand coins and stamps, the latter of which were adorned with profiles of each member of the Bates family, when one of the workers, clutching a pair of binoculars, hollered, 'Intruders! They're on the VHF radio. They want Prince Michael and Ben Fogle to come out on deck.'

Blimey, I thought, I've been rumbled by the secret services for consorting with an enemy of the state. A little nervously, I walked out on to the deck, which was riddled with holes, through which white water could be seen lapping at the fort's stumpy legs. I peered over the fort's rust-eroded banister and saw a small power boat bobbing about in the swell below. A man in a donkey jacket with a loud-hailer appeared.

This is it: the Germans are back, I thought, waiting for the words: 'Throw down your weapons and put your hands in the air!'

Instead I heard the following: 'Hi, I'm Danny Wallace from the BBC. Can we talk to Prince Michael, and could Ben Fogle please hide inside because we don't want him in shot?' BBC? Danny Wallace? Broadcaster and author of *Join Me* and sometime presenter of the paranormal on *Richard and Judy*? What the hell was going on? I had just spent nearly a year waiting for permission to visit Sealand and suddenly the BBC, together with a rival author, show up all at once.

'We had *Loaded* magazine here last week,' said Prince

James, peering below. 'And *Czech Penthouse*. And we were on the front of *Wired* magazine, just like in the film.' He straightened his collar, adjusted his hair and boarded the wooden rope seat that was once again dangling precipitously over the raging surf below, his gold Rolex watch glinting in the late morning sun as he descended to the tiny boat.

With Prince Michael otherwise engaged, the young princes offered me a guided tour of their country. The main quarters were set out along the Row, a corridor off which the day-to-day administrative headquarters were based. Next to the kitchen there was a living room with a small porthole window, sparsely furnished with two old sofas that had certainly seen better years, an old television and VCR and a vast collection of videos. Next was the royal bathroom. Much of the floor had corroded, leaving gaping holes to the surf below. This certainly wasn't a loo for sufferers of vertigo, or for royalty for that matter.

The upper deck is the dominating feature of Sealand, but the chunky legs hide the cavernous world where HavenCo had set up its HQ and operating room. Each block – one the North Sea Deep, the other South Sea Deep – has eight floors leading to the freezing seabed 100 feet below. A small hatch and a ladder let us down to the first floor of South Sea Deep, or 'biscuit' as the levels were called by the soldiers stationed there during the war. Level One used to be Prince Roy and Princess Joan's state room but had recently been converted into the nerve centre of HavenCo. I saw the dozens and dozens of computer hard drives and storage facilities cooled by fans and air conditioners that had been neatly stacked around the room. In the corner there was a discreet screen for 'monitoring'. 'It's where we check for "flesh",' explained Mike the mechanic, referring to their zero tolerance of the wrong kind of porn.

Level Two is now the fort's telephone exchange, which receives data from around the world using fibre optics, satellites and microwaves. With its dozens of laptops with no one sitting at them, it felt distinctly eerie, rather like an office out of hours.

'So is it busy?' I asked.

Prince Liam shrugged. 'Not quite as busy as they had hoped.'

Most of North Sea Deep had been designated as the nation's sleeping quarters, but on Level One there was a generator room, and on Level Five a conference room. A notice on the wall carried the instruction that the room must be booked ahead to avoid double bookings. The last, and only, entry in the diary was for 12 June 2002, booked to HavenCo for a 'planning meeting'. I was rapidly getting the impression that business was looking a little gloomy for HavenCo.

The lower I descended, the darker and colder it became, and soon I could hear the sea lapping against the three-foot-thick walls. On the furthest level down, before a sheer drop to the base of the leg, was the brig, a dark cavernous room with a thick steel door, behind which Gernot Putz had spent his six weeks in captivity.

This had once been a magazine room, where ammunition had been stored. I could hear the clanking of Channel marker buoys as they dragged their heavy chains across the sea floor. I could even make out the underwater echo of cross-Channel ferries, their engines giving out a low dull roar that reverberated around the damp room.

At one time, during the war there were as many as 300 soldiers stationed on this tiny fort. Some were here for weeks at a time, rarely seeing daylight. It certainly wasn't a popular posting. There have been several attempts by the government to requisition the old fort. The Royal Navy turned up once with a handful of military policemen claiming that Prince Roy had sold the fort. Young Prince Michael, fifteen at the time, was the only person on Sealand. A couple of Royal Marines started to climb a ladder and, in the spirit of a gallant prince from times gone by, Michael produced a pistol and the soldiers soon scarpered into the distance. Recently declassified documents have revealed the full extent of the government's worries about Sealand, including an order from Harold Macmillan's government to take the fort by force, but without the use of guns. Apparently, the Prime Minister was worried about the impact a fatality of a British soldier might have on the British public. To this day the Bates family remain certain that, should they ever leave the fort unattended, it would be blown up at the first opportunity by the British authorities. That may be a slight exaggeration, but Sealand has certainly proved to be a thorn in the side of successive British governments.

I asked the princes how they felt about taking on the government of Sealand, which has been continually occupied since their grandfather 'took' it back in the sixties. They shrugged in unison. 'I'd like to manage Sealand's football team,' said the youngest, Prince Liam. It seemed amazing that an island struggling to find the manpower to stay viable had enough people for a ping-pong tournament let alone an international football team, and yet a team genuinely exists. 'They've already played one international in Denmark,' the young prince said, beaming. 'They drew two all against Checkia,' interrupted Prince James, 'and they'll be taking on Aaland in the spring.' I wondered where on Sealand they trained.

It turns out that Sealand even has an international athlete, a certain Darren Blackburn, originally from Canada, who appears in various events from time to time, although the exact nature of these meetings I was unable to establish. 'The Sealand flag has even been flown from Everest,' Prince James informed me as we headed back towards the deck. It felt good to return to the surface. Sealand certainly isn't a country for the claustrophobic.

Sealand is just one speck in a post-modern archipelago of real and virtual micro-nations or mini-states born of puckish claims by free-spirited pioneers, adventurers and of course con artists. One website lists more than fifty entities with names such as Redonda,

Oceanus, Vikingland, Elgaland-Vargaland and the Dominion of Melchizedek. Only recently New Yorker Gregory Green founded the New Free State of Caroline, a nation state based on a speck of coral in the middle of the Pacific Ocean. Green spent hours scouring maps for as-yet unclaimed territory. The island of Caroline appeared to meet all the criteria required for starting a new nation: it was uninhabited, appeared to have been so for quite a long time and had been claimed over the years by five different countries. Green threw his hat in the ring.

He wrote letters notifying the United Nations (UN) and various other international bodies of his intentions for Caroline. He recruited 2,000 citizens. He designed a flag, established embassies in major world cities and set up a pirate radio station to give his citizens access to the airwaves. Unfortunately for Green, he later learned of archaeological finds showing that prehistoric people once lived on Caroline. According to international law, that gave Kiribati, a nation of thirty-three coral reefs inhabited by the descendants of ancient Pacific islanders, firm possession of the island. For Green and thousands of others who dream of owning their own country, Sealand remains the role model and arguably the original micro-nation.

Michael and his father have made every effort to have Sealand recognized as a sovereign state. In 1987 Britain passed a law extending its territorial waters from three miles to twelve, but the day before the law came into effect, Sealand announced its own territorial waters to be twelve miles. Sealand and the UK appear to have reached an impasse. The British refuse to comment on Sealand, let alone take it by force, as this would in effect be admitting its existence. Lew, meanwhile, continues to work on Sealand's authentification, even submitting each year's imports and exports and balance of payments to the UN. He argues that the court cases in the UK and the German fiasco of 1978 amount to de facto acknowledgement of Sealand's sovereignty.

On Sealand truth really is more fantastic than fiction.

As we prepared to leave his sovereign land, Prince Michael proudly held up a copy of the *Hollywood News* with a front-page story about Sealand. 'They're going to make a film about us,' he said, beaming with his roguish charm. (The Prince was certainly a very likeable character.) According to the article, Warner Brothers had bought the film rights to Sealand, and studio bosses had already lined up Russell Crowe to play Prince Michael. 'Mel Gibson would be better,' said Michael, laughing, 'but they'll probably end up casting Christopher Biggins.'

# Rockall

I had been illuminated by Sealand, not by Prince Michael's rusting fort, but by the idea of not just moving to an island but of owning my own unique sovereignty, my own country. I needed a place that was not only uninhabited but also constitutionally challenged. It would need to be remote yet attainable, and rugged but charming. Throughout my journey offshore I had heard tales of such a place, an uninhabited island disputed by Britain, Ireland, Iceland and Denmark, an isle rich in minerals and, as I would soon discover, astonishingly rich in history: Rockall, a storm-battered rock famous for its appearance in BBC Radio 4's shipping forecast and visited by only a handful of people. If I stood upon her summit she would be mine. The problem, however, would be getting there.

Rockall sits alone, over two hundred miles from the nearest land, the mere peak of an ocean mountain, measuring just 110 feet in diameter and 63 feet high. Given its size and location, in the North Atlantic Ocean about three hundred miles from the coasts of Scotland, Ireland and Iceland, this tiny nugget has enjoyed a surprisingly colourful history. To this day it remains at the centre of one of the world's most unusual and bitter custody battles. Battered by some

of the heaviest seas on the planet, with barely a hint
of vegetation upon her, let alone a dwelling of any
kind, this inhospitable rock may not appear to be a
prime piece of real estate; but for those, like myself,
bent on owning an island they can call their own,
Rockall offers temptation beyond endurance. I am,
however, just one of its many suitors.

Rockall is technically the most westerly point in
the United Kingdom, but its status has long been
disputed. The island itself is of little consequence to
those who claim sovereignty over her: it is what lies
beneath. Money, of course, lies at the heart of the
matter, because the area round her is reputed to be
rich in petroleum, while her waters teem with valuable,
increasingly rare fish stocks.

Rockall is by no means alone in having her ownership disputed. There is a long history of such territorial
squabbles around the world, particularly where natural
resources are involved; many of them are going on as
I write. The fruits of the sea have turned many islands
into pawns in an international game of world chess.
There have been major international crises over disputed territories, such as that between Argentina and
the UK over the Falkland Islands, as well as smaller
tiffs such as the ongoing row between Spain and
Morocco over Parsley Island, an uninhabitable rock in
the Mediterranean that recently became the site of
the first European military invasion since the Second
World War. In July 2002, during a secret operation,
Moroccan soldiers took the Spanish-owned island,

which lies just 650 feet from their coastline. Incensed by this effrontery, the Spanish responded by sending half-a-dozen warships to the rock, which was eventually liberated by a team of elite Spanish assault troops who captured the occupying force and lowered the North African red and green flag. The standoff ended peacefully, but not without the intervention of NATO.

The South China Seas have been the stage for numerous disputes in recent times. Japan is currently involved in a face-off with China over two tiny uninhabited outcrops of reef 1,000 miles south of Tokyo and has spent £130 million on shoring them up to island status, so as to ensure access to the rich surrounding seabed. There is another dispute over an archipelago north-east of Taiwan, regarding seabed gas exploration rights. In 1995 the People's Republic of China occupied the fantastically named Mischief Reef, creating a substantial crisis in South East Asia to add to the quarrel over the Paracel Islands, which are claimed by China, Taiwan, Brunei, Malaysia and the Philippines.

Even Canada and Denmark, two of the most eminently sensible nations on the planet, have been at each other's throats over a dot on the map known as Hans Island, which is even more remote and inhospitable than Rockall. On the face of it, the two nations are fighting over less than a square mile of useless land but, like the Rockall imbroglio, the dispute is not so much about the island itself as about its surroundings – in this case, one of the last virgin territories on earth, the North Pole. A tiny island wedged between Greenland, a semi-autonomous Danish territory and Canada, Hans Island has found itself in the thick of a scramble for the Arctic. In 1973, the two countries established a border down the inhospitable Nares Strait halfway between Greenland and Canada's Ellesmere Island, but the two nations decided that sovereignty of Hans Island, 683 miles from the North Pole, would be decided later.

When Canada's defence minister set foot on the island and raised Canada's maple leaf flag, relations between the two nations deteriorated. The Danes responded by sending in their Arctic patrol cutter. They argued that the area is a natural extension of Greenland; therefore the North Pole is part of Denmark, and the rights to exploit its potential wealth of oil and minerals are Danish. The Hans conflict is currently being fought in cyberspace as hundreds of bloggers from both sides have hijacked the web with patriotic hyperbole. My favourite entry to date comes from an incensed Dane and reads simply: 'Hans Island – does it sound Canadian?'

The Danes also have designs on Rockall, but I'm afraid they'll have to join the queue along with the rest of us. According to legend, when Finn MacCool, the legendary giant of Irish folklore, flung a sod of grass and a pebble across the Atlantic in a fit of rage at a fleeing rival and the sod took root to become the Isle of Man, the pebble formed Rockall. (Another legend suggests that Rockall is the extremity of the ancient kingdom of Brazil. Scientists, meanwhile, believe that it is the tip of a volcano that last erupted fifty million years ago.) While it is unlikely that any other states will go to the international courts of justice to settle the dispute, the Irish continue to question the UK's possession. Rockall is seen by some as the tip of a Republican iceberg, as the militant song 'Rock on Rockall' makes clear, with its use of the convenient rhyming potential of Rockall, Whitehall and Donegal. The Wolfe Tones, an Irish Republican band, even penned a song about Rockall (words by B. Warfield):

Oh the Empire it is finished
No foreign lands to seize
So the greedy eye of England
Is turning towards the seas
Two hundred miles from Donegal
There's a place that's called Rockall
And the groping hands of Whitehall
Are grabbing at its walls . . .

Now the seas will not be silent
While Britannia grabs the waves
And remember that the Irish
Will no longer be your slaves,
And remember that Britannia – well
She rules the waves no more
So keep your hands off Rockall –
It's Irish to the core.

Icelanders have been just as indignant about Britain's appropriation of the island but rather more subtle in their approach to musical propaganda. The Icelandic jazz funk band Mezzoforte recorded an instrumental track called 'Rockall' which was used in the BBC Radio 1 chart countdown during the eighties. In 1998, one Irish opponent, the former Mayor of Dublin, Séan Dublin Bay Rockall Loftus, argued that the chunk of rock was nineteen miles closer to Ireland than to the UK and that the British should relinquish control. The *Economist* described the claim as 'as scientifically unimaginative as it was legally impeccable' – the Irish had simply drawn a line from coast to coast and claimed everything in between. The former mayor even changed his name by deed pole to 'Rockall' to show his support for the Irish cause.

Ireland has also attempted to validate her claims by occupying the small island. In 1992 Philip and Fergus Gribbon, brothers and businessmen from County Donegal, set out on an expedition. They planned to paint the Irish tricolour on one of Rockall's cliff faces, but their trawler's engine broke down halfway and they were forced to abort their invasion.

The tiny rock has only recently been mapped accurately and is still absent from many maps. Perhaps unsurprisingly, it has been responsible for a number of wrecks through the years, the first recorded being in 1686, with the loss of 250 lives. It was only discovered when a boatful of bearded Frenchmen rowed several hundred kilometres to St Kilda, from whence they were eventually rescued. The first written mention of the rock is on a Portuguese map dated 1550, but even after that, with its guano-covered summit and its unique shape, the rock was often mistaken for a ship in sail.

The first known landing was in 1811 by Lieutenant Basil Hall, who was aboard HMS *Endymion* while she was patrolling the waters around Northern Ireland and spotted the rock, thinking at first, like many before, it was a ship. Hall described it in his diary as 'a solid block of granite growing out of the sea at a greater distance from the mainland, than I believe, any other island, isle or rock of the same diminutive size to be found in the world'. Intrigued, he sailed close and circled the strange isle before mustering a landing party, who became in all probability the first people to set foot on the tiny isle. Unfortunately, shortly after ascending the vertiginous rock the landing party became stormbound: the effects of the giant swell of the sea were compounded by a thick fog that obscured Hall and his group from the *Endymion*. They were forced to spend a night on the rock before a rescue party was able to locate and rescue them. To this day the only flat area of the rock is known as Hall's Ledge. If only I had been around in 1811.

The next recorded visit to the island was some fifty years later in 1862, by *Porcupine*, while she was laying underwater cables for the first transatlantic telephone cable. Thereafter the volume of shipping around Rockall escalated, as trawlers were attracted by its bountiful fishing, and at the turn of the century the rock suffered its greatest tragedy when the *Norge*, a 3,000-ton liner en route from Copenhagen to New York, foundered on its hidden reef, with the loss of more than 600 lives. For the first time there were calls for a lighthouse to be built on Rockall to prevent further loss of life in her volatile waters.

My plan was simple: sail to Rockall, invade, raise my flag and declare myself King. Letters would then be sent to the relevant governments alerting them to this occupation and, Bob's your uncle, it would be mine. Simple really: all I had to do was get there – or so I imagined.

There are plenty of precedents for my land grab, and my plan wasn't particularly audacious compared with others. Some have even tried to claim parts of the universe on the basis of the Outer Space Treaty of 1967, which states that 'the moon and other celestial bodies, are not subject to national appropriation by claim of sovereignty.' Gregory Nemitz went to court to try to recover money from NASA for a 'parking fine', claiming that the organization had unlawfully landed its probes on his private property. The land in question was Eros, one of the largest asteroids near Earth. Mr Nemitz was unsuccessful.

My favourite story is that of Lester Hemingway, the eccentric brother of Ernest. He set off on a raft of bottles from the coast of Florida and declared himself leader of the republic of Atlantis, but not even a handwritten letter from the then President Johnson addressed to the leader of Atlantis could help his cause and the republic ended before it had begun. A little like my first attempt to land on Rockall, described at the start of this book.

On that occasion I had returned to the mainland, my spirits as damaged as much as the trawler that had tried so hard to take me there. My island dream had been scuppered by the weather, but I was determined to make another attempt at the first realistic opportunity. I wasn't about to give up that easily, but I knew the pressure was on. My plans for that first assault had been thrown into jeopardy when I received a call from a friend in Scotland, who said, 'I think you should see the papers.' The Scottish press had got hold of the story of another man's attempt to claim Rockall. It transpired that my rival was writer Charlie Connelly, author of the fine *Attention All Shipping*, who let slip at a literary lunch his plans to take the remote rock. I couldn't believe it: Rockall had remained largely forgotten and untouched for nearly ten years and now suddenly two people had decided to set out at the same time. It's like buses, I thought. A few carefully placed calls and a little research had revealed that Connelly was planning his attempted coup in June. I still had several months on him, and unperturbed, I had forged on with my own attempt in March, only for Mother Nature at her angriest to foil my plan – much to the glee of several newspapers. 'Fogle's Folly Foiled' ran one headline.

Connelly and I, though, were not the only competi-

tors in the race, it seemed. 'Dear Mr Fogle,' read an
e-mail in my inbox. 'We've been following the story
of your attempt to land on Rockall. We made an
unsuccessful attempt to land on the sacred isle in 2003
and will be setting out again in June to finally claim
it as ours. Sorry but 'Benland' will have to remain a
dream as we intend to declare the island a people's
republic.' The e-mail was signed by the editor of the
*Rockall Times*, a very surreal website dedicated to the
island with the slogan 'There's f**k all on Rockall.' A
postscript read, 'Rockall in March? Madness, but we
salute your intrepid team.' I was perplexed. By their
own admission there was f-all there, so why were they
so keen? Whatever their motivation, it now appeared
that a third party had joined the fray. Lester Haines,
the editor, and his team of 'cyberhacks' were planning
an invasion of their own, setting out from Stromness
in the Orkneys in June. Rockall was suddenly a very
popular place. At this rate, I thought, Butlins was
probably drawing up plans for a North Atlantic theme
park and Easyjet was probably costing daily flights. It
wasn't the only e-mail in my inbox. Charlie Connelly,
too, had heard about my attempt: 'I look forward to a
good old no-holds-barred, handbags-at-dawn battle for
sovereignty.' He had thrown down the gauntlet. The
race for Rockall was on.

June is Rockall's weather window and it now seemed
that three expeditions would all be setting off together.
What would I do if another flag was already flying
from its peak when I arrived? Was it first come, first

served? What happened if I was invaded? But these questions would have to wait, as I still had to work out how to get there.

During the Cold War, a number of islands around the UK, including the Hebrides, were used as firing and training ranges. South Uist had become a particularly sensitive location, and Harold Macmillan, then Defence Secretary, feared that the Russians might seize Rockall and use it as a base for spying on the Hebrides. Quite how the Cabinet envisaged the Russians occupying a rock barely large enough for a gull's nest, let alone a listening station, is hard to imagine, but the fear prompted the British to formally claim the rock as theirs. On 14 September 1955, HMS *Vidal* set sail for Rockall, from which two Royal Marines and a builder were lowered on to the storm-ravaged rock by helicopter in what was arguably the last act of territorial expansion by the British Empire. The newest addition to Her Majesty's 'overseas collection' was baptized with the raising of the Union flag and the cementing of a plaque, which read:

By authority of Her Majesty Queen Elizabeth the second, by the Grace of God of the United Kingdom of Great Britain and Northern Ireland and her other realms and territories, Queen, Head of the Commonwealth, Defender of the Faith [etc, etc, etc . . . ] and in accordance with Her Majesty's instructions dated the fourteenth day of September One thousand, Nine Hundred and Fifty-Five, a landing was effected this day upon this island of Rockall from HMS *Vidal*. The Union Flag was hoisted and possession of the island was taken in the name of Her Majesty.

R.H. Connell. Captain, HMS *Vidal*, 18th September 1955

As the flag was raised, Lieutenant Commander Scott announced, 'In the name of Her Majesty Queen Elizabeth II, I hereby take possession of the island of Rockall.' HMS *Vidal* fired a twenty-one-gun salute and the team was tasked with installing the rock's first light, to prevent another *Norge* and no doubt to justify the vast expense to the taxpayers of this imperialistic expansion. Photographs of the takeover were syndicated worldwide by the government's spin doctors to reinforce the claim. Thus Rockall became a British outpost in the North Atlantic, an isle that could challenge the Falklands and even South Georgia in the inhospitality stakes.

All was quiet on Rockall's western front until 1971, when Denmark, Iceland and Ireland suddenly challenged the sovereignty, not for the lump of guano-smothered granite itself but for the vast reserves of oil and gas believed to lie and bubble beneath. Britain responded with the Rockall Act of 1972, which officially made Rockall part of Inverness-shire (which was bound to terrify the claimants), and imposed a fifty-mile exclusion zone around the rock. The government even began issuing petrol-drilling rights and licences, and then to solidify their claim it flew two bearskinned

Royal Marines in full ceremonial dress and a sentry box out to the island at the taxpayers' expense, lowered them on to the rock and photographed them, thereby asserting that it had to be ours because it had a sentry box and two Marines. One assumes it wasn't large enough for a red phone box or a tandoori.

The responses to these acts were as mixed and peculiar as Britain's own actions. J. Arbach Mackay insisted that the Admiralty hand Rockall back to the Mackay clan, as his father had claimed the island back in 1846, and, while the government understandably ignored his claim, the musical satirists Flanders and Swann felt compelled to pen the following:

> The fleet set sail for Rockall,
> Rockall,
> Rockall,
> To free the isle of Rockall
> From fear of foreign foe.
> We sped across the planet
> To find the lump of granite
> One rather startled gannet;
> In fact we found Rockall . . .

While the British were bemused, others were infuriated. In April 1973 the Rockall Liberation Front was created. This was an underground network based in Iceland whose political wing continue to write letters of protest every week to the British embassy in Reykjavik. Denmark and Iceland reinforced their claim by arguing that Rockall was uninhabitable and therefore couldn't be called an island – in which case the British composition of an exclusion zone would be invalid. In 1982 Britain signed the United Nations' convention on the law of the sea, which concluded that 'rocks which cannot sustain human habitation or economic life of their own shall have no exclusive economic zone or continental shelf.' So in 1985, in the best traditions of British eccentricity, the British government sent SAS man Tom McLean to the island, where he lived in a specially constructed pod for a little over a month. McLean certainly had all the qualifications for the job. In 1969 he had become the first man to row solo across the Atlantic, and in 1985 he had added the record for the smallest boat crossing after traversing the ocean in a craft just seven feet long. He made Rockall his home for an impressive thirty-nine days, relentlessly battered by the weather, which forced him to take shelter for most of his stay in his tiny three-foot-high hut which had been anchored to the summit. He survived on a diet of tinned turkey and long-life milk.

By living on Rockall, McLean not only proved it was habitable and therefore an island, not a rock, but also reinforced the UK claim to sovereignty. The UK was rather pleased with its efforts, and Rockall suddenly found itself in the spotlight again.

After Tom McLean's impressive tenancy all was quiet on the Rockall front until 1997. Greenpeace staged a daring coup and the island hit the headlines once again. Three activists landed on the rock with a solar survival capsule and occupied it for a staggering forty-two days, living off rainwater and freeze-dried food. Using Britain's own criteria Pete, Al and Meike, Greenpeace's three watery 'Swampies', argued that as they had stayed on the rock for longer than anyone else, that made Rockall theirs. They asserted the claim by renaming the rock Waveland and declaring it a global state committed to worldwide environmental protection, announcing: 'By seizing Rockall we claim her seas for the planet and all its people and no one has the right to unleash this oil on our threatened climate.' They offered citizenship to anyone prepared to take the pledge: 'without violence and by bearing witness, to defend nature, to protect the global commons, to reform industrialism, and to secure peace, believing in action rather than words.'

Much to the astonishment of Greenpeace, the government conceded that 'Rockall is a British territory. It's part of Scotland and anyone is free to go there and can stay as long as they please.' Greenpeace's call for citizens to move to Waveland went unheeded and eventually Waveland disappeared below the waves like Atlantis.

My task, second time round, was to find a more reliable way of getting to the remote outpost. I spoke to a helicopter company in Inverness that made sorties around the Outer Hebrides, only to discover that single-engine choppers aren't licensed to cross the oceans, while twin-engine helicopters can't manage the return trip without refuelling – something glaringly impossible in the vast watery desert of the North Atlantic. I doubted the Royal Navy would be helpful in my revolutionary and distinctly unpatriotic, possibly treasonous, endeavour. I was worried they might put a stop to it if they discovered my secret plan. My suggestion to the Stornoway coastguard of using a trip to Rockall aboard their Sea King helicopter as a training exercise was politely declined with much amusement and a warning about the dangers of such a journey.

I called a friend, Angus MacDonald, on Uist in the Hebrides whom I knew had a RIB (rigid inflatable boat) that he sometimes took to St Kilda, a round journey of nearly a hundred miles. I suggested we make the 300-mile round trip in his small boat. 'To Rockall?' he repeated incredulously. 'In an open boat?' He didn't need to say more. I spoke to the owners of trawlers and dive boats, but they were busy, fully booked, too expensive, unable or simply unwilling to make the long treacherous journey. And then I remembered the *Eda Frandsen*, a 1939 Danish fishing boat converted by her owner and skipper Jamie Robinson into the magnificent sixty-five-ton, fifty-five-foot timber gaff cutter she is today, on which I had once spent a week sailing around the small isles of Scotland. Jamie had bought the boat with his brother and converted it at their home in Doune, a remote west-coast peninsula where his family run a restaurant and hotel accessible only by boat.

They had spent a year meticulously working on the vessel and then just a week before it was due to be launched, a fire wrecked the boatyard and destroyed their boat. Nothing was left but her charred skeleton. The brothers dug deep financially and mentally and started all over again, offering free accommodation and food for any boat builders and carpenters willing to volunteer on her reconstruction. Much to Jamie's amazement, craftsmen flocked from all over the world to their remote peninsula for a chance to work on this beautiful cutter, and a year later she was launched. The *Eda Frandsen* was born of sweat, fire and tears.

Perhaps unsurprisingly, as he was from a family of restaurateurs, Jamie also happened to be an extraordinary cook, who managed to create the most fantastic recipes in the smallest of galleys and with the simplest of produce. He had entertained us with his accordion playing and old sea shanties and reassured me with his competent seamanship. With his slight belly and thick curly beard, he struck me as the archetypal salty sea dog. Jamie also knew the Scottish waters better than anyone, having had nearly thirty years' experience in navigating them. The problem was that summer was approaching, the *Eda Frandsen* was a popular boat, and she was often booked up years in advance for Jamie's popular whisky tours and 'gastronomic' holidays.

'Where?' hollered Jamie.

'Rockall!' I repeated down the fuzzy line.

'It's blowing a gale here,' bellowed Jamie above the wind. 'Where?'

'*Rockall!*' I screamed.

'Rockall?' he blurted back, as if I was mad. I was starting to get used to this sort of reply. 'Why do you want to go there? There's fuck all there!'

'Can you get me there, Jamie?' I begged.

'When do you want to go?' he asked.

'As soon as you think we can make it.'

'I'll get back to you,' he said.

The *Eda Frandsen* had been out of the water for the winter and wasn't due to be refitted and launched until May, for her first booking. She had been severely damaged in the winter storms by waves that had crashed over the breakwater and on to the jetty, and it would take at least two weeks to get her seaworthy again. A few days later Jamie called back: 'The good news is that we can make it. The bad news is we'll have to go next month.' April, I thought, was just a month after my ill-fated attempt the year before and still very much in the North Atlantic's winter, when gales pounded anything that dared to sail. Another problem was that even if we were able to reach the remote rock, there was a very real probability that we wouldn't be able to get anywhere near it, let alone on it, for fear of being wrecked on her treacherous reef. 'It's a long way out, but I think I can get you there, though I can't guarantee you'll get on to her,' he reiterated. Getting there was good enough for me; little technicalities like how to scale her could wait, I thought, as I made plans to go to Scotland. I was going back to Rockall.

Jamie had agreed to take part in the expedition partly on account of it being the only place he hadn't been to in his thirty years' skippering in Scottish waters. Given that Rockall isn't on the way to anywhere, that it's in the middle of nowhere and that it is often referred to as the stormiest place in the world, it is easy to understand why more people have been to the moon than to Rockall. Its name comes from Sgeir Rocail, meaning 'roaring rock' in Gaelic. William Golding made Rockall the setting for his 1956 novel *Pincher Martin*, a surreal and nightmarish vision of one man's battle for physical and mental survival in one of the most unforgiving and hostile environments.

Having chartered the *Eda Frandsen* for the voyage, I set out to find some fellow passengers to spread the cost and reduce the monotony of three weeks on the Atlantic. Perhaps unsurprisingly, my friends weren't particularly forthcoming when it came to doshing out for the chance to spend three weeks being seasick in the rain while their mate invaded a rock. But there was one man who I knew wouldn't turn down such an offer, Charles Veley, an American multi-millionaire who collected countries as most people collect stamps. I had met him first in Tristan da Cunha in the South Atlantic and then again in the British Indian Ocean Territories. He had started travelling obsessively five years ago, and had already visited 499 countries, earning an entry in the *Guinness Book of Records* as 'the world's most travelled man'. Having attained the title, Charles had then set about 'solidifying' his portfolio by visiting various disputed places around the globe. When I called him, he was in the Andaman Islands.

'Hi, Charles. Do you want to come to Rockall?'

'When?'

'In two weeks.'

'See you there.'

And that was it. Two weeks later, Charles landed in London from Somalia with a pair of shorts and a T-shirt, and we set off on the long journey to Rockall. We flew to Glasgow, where we picked up a hire car and drove for four hours to the picturesque Scottish fishing port of Mallaig, where the *Eda* was waiting to shuttle us to my island. It was still winter in the High-lands and the mountains were covered in an icing of snow. Jamie had managed to assemble an impressively eclectic crew for the expedition ahead. Tom was a thirty-year-old former member of the British bobsleigh team; then there was his girlfriend Anna, with a masters degree in lemurs; and George, a part-time postman from Cambridge, also in his thirties. Completing the cast was Beetle, a sprightly 76-year-old coleopterist (beetle expert to you and me) from the Natural History Museum in London and world authority on the click beetle.

Tom and Anna had joined the *Eda* for the season, taking the chance to spend the summer together sailing around the Western Isles. Tom was the relief skipper and Anna was cook, while George was crew, but I couldn't work out where Beetle came into the arrange-ments. I later discovered that she lived aboard the *Eda*, in much the same way that a woman lives aboard the *QEII*. Beetle had discovered sailing rather late, in her seventies to be precise, but had become a fanatical devotee and spent as much time as possible on the high seas. She books a berth for the whole season, sailing wherever the boat takes her; Beetle was part of the *Eda*'s furniture. It was certainly a diverse invasion force.

In 1994, there was a most unusual exchange in the House of Lords during a debate about the policy of North Atlantic oil exploration and the status of Rockall. Lord Campbell of Croy reminded the house about Tom McLean's patriotic act, which had solidified the UK's claim to sovereignty of the rock: 'My Lords, experts in international law have pointed out that the UK position on Rockall might be strengthened if there were to be a British wedding, or better still, a birth on Rockall.'

Earl Ferrers: 'My Lords. I am not quite sure what one concludes from that, other than the fact that any-one who tried to be married on Rockall must need their head examined, as must anyone who tries to have a baby there.'

Their exchange caused widespread chuckles across the benches, but it got me thinking. Ticking off 'first' boxes for Rockall could only strengthen my claim. I would have the first international athlete among my boarding party, as well as an American world-record holder. It seemed a safe enough bet that a lemur expert had never been there before either, let alone a coleopterist. I felt certain that Beetle would rate as the oldest person to set foot on the rock, and if I could somehow persuade Tom and Anna to procreate on the island that would be the icing on the cake.

The *Eda* was a striking boat with her 100-foot mast whittled from a single Canadian oak, her teak decks and her handmade crimson canvas sails. Her bulging green waist sat close to the water, weighted down by her 65 tons, which meant that even the smallest wave or swell crashed over her belly, sending torrents of water down her deck and on to those who sailed her. 'You'll need these,' announced Jamie, handing us each a pair of yellow wellies. 'She's designed to get wet.' He smiled knowingly. Charles had arrived worryingly unprepared for the stormy arctic conditions of the North Atlantic, but some frantic last-minute phone calls had procured us a mountain of wet-weather gear, fleecy leggings, fleecy vests, fleece trousers, fleece jackets, waterproof salopettes and some state-of-the-art ocean jackets complete with integrated survival systems – all in dazzlingly bright colours, of course.

Charles and I were assigned our bunks up in the foc's'le, a tiny cabin with twin bunks next to the anchor chain and barely large enough for all our kit, let alone us. I struggled into my multiple layers, contorting myself with each movement, banging elbows and knees as I fought my way into the unfamiliar garments. I could hear cursing from Charles as he manoeuvred into his alien spacewear. Fully kitted, I trundled through the galley and up the steep steps and squeezed my way through the hatch on to the deck. Feeling like a fat duck as I waddled along in my eight layers, I was slightly disconcerted to find Jamie in only a pair of waterproof trousers and a tatty old T-shirt. My chameleon-like transformation from city dweller to ocean master was rather less than perfect, as Jamie pointed out: I had the layers on the wrong way and my salopettes were back to front. This was going to be interesting, I thought, as I set about redressing.

This comic picture of incompetence was completed by Charles, who lumbered past like the Michelin man, labels and tags still dangling from his jacket. All the gear, no idea, I thought, as we ambled around the unfamiliar deck in our neon-yellow jackets, while the rest of the crew buzzed around in their heavily weathered and sun-bleached foulies. I was particularly amused to see George dressed from head to toe in full Royal-Mail-issue jacket and trousers. I half-expected him to hand me a pile of letters as he scurried around the deck preparing the *Eda* for her departure.

The weather wasn't in our favour, but Jamie had estimated that it would blow itself out by the time we crossed the Hebrides and headed out into the Atlantic. By his calculations it would take us five days to reach Rockall. We would then have a few days to wait for a break in the weather during which we could make our assault. It was dusk when we finally nosed out of Mallaig and into the gale-force winds of the Sea of the Hebrides. The ocean had been stirred into a white frenzy by the relentless wind. Cloud obscured the mountain peaks of Skye and Rum across the Minch, which separates the Outer Hebrides from mainland Scotland. It was an inauspicious start, frighteningly similar to my last attempt's end. The sky was streaked with grey cloud and intermittent squalls chilled us with tiny hailstones that clung to us for maximum effect. My teeth chattered uncontrollably; my hands were blue, my nose scarlet. It might have been April, but there was certainly no sign of spring.

Storm-lashed, we rolled and lurched our way across the Minch and through the relative calm and tranquillity of the Sound of Barra, a natural sea waterway protected by steep mountains on both sides, before reaching the mighty Atlantic and its powerful waves which had travelled and grown for nearly three thousand miles before depositing their watery load on us. Water cascaded over the deck, swamping us up to our waists. I realized why Jamie had issued us with wellies and, more alarmingly, a harness, without which we would surely have been washed overboard.

For two days and an interminable night we fought our way across the stormy ocean deluged by rain, sleet and snow. My body struggled to adapt to the unnatural movement and even my eight layers couldn't protect me from the glacial conditions, the damp icy air chilling me to the bone. At last land appeared on the horizon: not Rockall, but the land nearest her, St Kilda, more than a hundred miles south-east of our destination but providing a welcome anchorage where we could take a break from the incessant and icy headwind. We were only a quarter of the way to a place that, according to the Admiralty books, never has a calm day. We pulled into the relative safety of her bay and Jamie ferried us ashore.

'Rockall?' chuckled the barman. 'In April?' Laughter engulfed the island's famous pub, the Puff Inn. We were the first boat of the season to reach St Kilda. It would be another month before the tourist season began in earnest. A wall of the Puff Inn was adorned with a huge white banner proclaiming 'The Great British Rockall Expedition', a memento from Tom McLean's 1980s expedition. I wondered whether I would return here too for a celebratory drink, and prop up the bar with tales of my derring-do.

We bade farewell, collected our last shipping forecast, which promised a break in the weather, and sailed away from the 'edge of the world' and back out into the bitter Atlantic wind, towards the land that I was determined to claim as my own. For twenty-four hours we sailed across the tumultuous ocean, working in three-hour shifts. Sleep was nigh impossible as we reeled and rocked our way across the winter seas. We were banking on a window opening for us: our hopes rested on a low-pressure zone in the area being occluded by two high-pressure ridges that would bring calmer weather. But instead the window appeared to be closing, as both the size of the ocean and the power of the wind were increasing. We battled on into the eye of the storm until the early hours of the fifth day. With such a small crew, we were divided into two watches of three. We were now each spending four hours at the helm, followed by four hours of sleep. Day became night and night day. My body clock struggled with its new routine.

'Grub's up!' hollered Jamie from the steamy galley. Incredibly in the space where I had struggled to lie down, Jamie had somehow created a banquet: a full no holds barred, all the trimmings, full monty, all singing all dancing, bells and whistles, good old British Indian curry. He produced fresh chutneys, yoghurt and poppadoms and even provided a choice of hot or very hot tikka masala. Somehow, while the boat lurched and pitched and generally bucked like a theme-park ride, Jamie had managed to present a feast that would be the envy of the finest curry houses. How he managed it I'll never know; eating it was difficult enough. I chased the basmati rice around the cabin, while the chutney appeared to chase me. The smell of the curry soon began to overwhelm my senses and clog my pores with its thick pungent aroma. It suddenly felt very stuffy in the cabin. Sweat spread across my forehead. I nibbled on a poppadom while Jamie tucked in. As I turned the same colour as the apple chutney, I realized why curry is not normally recommended as a cure for seasickness. Moments later I made my excuses and regurgitated my poppadom.

We had been sailing for nearly a week now and the days had become a blur as I slipped into an hallucinogenic reverie, fuelled by vast quantities of seasickness tablets. As I lay in my bunk for my four hours' reprieve from the cold driving rain, the heavy waves thumped against the bow, the sound reverberating around my head and denying me sleep. Every so often a rogue wave would catapult me from the top bunk to the ceiling, before returning me to the floor. I lay in this state, barely half asleep, until someone shook me from my trance and I crawled back out into the arctic conditions for my watch. Day in, day out, we continued the routine of four hours on and four hours off, the monotony broken only by an occasional cup of hot water and half a poppadom.

Sleep-deprived, cold and very, very seasick, I found myself at the helm. It was two o'clock in the morning, but heaven only knows what day it was. The wind lashed around us, whipping the sea into a seething cauldron. Mountainous waves towered over us, crashing over the bow, and white water raced down the deck, drenching us with each pitch and fall. My safety harness kept me from being washed over the side as water swamped the aft deck. For several hours rain and sleet hailed down as we beat into the unending blackness, my helming broken intermittently by an empty retch as my contorted stomach told me to end this ordeal.

'It's not good,' shouted Jamie as the freezing rain and hail lashed our faces and water swamped us up to our knees. I was experiencing a sense of déjà vu. Jamie had called the Met Office on the satellite phone to be told that we were in a force ten and Rockall in a force twelve. The *Eda* swung like a pendulum, her 100-foot mast, gaining momentum with each wave, rocking us uncomfortably from side to side while errant waves knocked us back and forth. 'There's no visibility,' explained Jamie. 'We won't get anywhere near it. The swell is forty feet out there.' Even fishing boats had retreated to the safety of harbour.

With heavy hearts and lightened stomachs we ceded victory to the elements and turned around, sixty miles short of Rockall, and headed for home. Once again my carefully drawn-up plans had been crushed by the weather. It had all been for nothing. It was enough to make me weep.

Our battered little boat beat into the wind back towards the relative calm of the Outer Hebrides, once again passing St Kilda, now an invisible blot on the horizon. For three days we punched the heavy swell, as the storm passed overhead on its way to Canada, to be replaced by sunshine which illuminated an ocean still bubbling with anger. Though the swell had already started to subside, Jamie suggested we break the return journey with a night's shelter in the Monarch Isles to allow the sea even more time to calm. I certainly wasn't going to argue.

About five miles west of North Uist in the Outer Hebrides, the Monarchs are a collection of little more than sand dunes with a rocky base, rising no more than sixty feet above the Atlantic waves. Though the islands were inhabited before AD 1,000, when they had a population of over a hundred, like many of the Scottish islands, they were abandoned in 1931 when landowners discovered it was more economical to graze sheep on their land than house people; only a lone lighthouse keeper remained, until automation in 1942. Currently the islands, a National Nature Reserve and European Special Area of Conservation, are home to 100,000 grey seals, the second largest colony in the world. They were also the perfect place to break from the incessant Atlantic onslaught, as the presence of the seals testified.

It isn't always calm here, of course. The great rusting hulls of ships wrecked against the treacherous reef and littering the islands are evidence of the ocean's ferocity. A huge brick lighthouse dominates the low-lying islands, abandoned and crumbling like a gravestone to the wrecks.

'DO NOT ENTER' tempted a warning sign on the lighthouse door, firmly closed with a padlock, though not so firmly that a carefully placed screwdriver couldn't open it. I thanked my scout training and like the Famous Five we entered the dark tower. Wood pigeons, startled by our presence, escaped through the open door in a cacophony of surprise at the intruders entering their home. I felt like a naughty schoolboy as we ascended the spiral staircase of the pitch-black tower with a small Maglite to illuminate our way. Up and up we slogged and wheezed, the seven days at sea having taken their toll.

A bright sun beat through the glass panes, casting eerie shadows across the rusting light, now sadly extinguished. We climbed out on to the small parapet and the *Eda* looked tiny as she bobbed up and down in the turquoise waters. I could make out the mountains of Lewis to the north-east, while to the west the ocean spread out like a giant grey quilt as far as the eye could see. It was a welcome distraction from the week's storms and a chance to forget about the disappointment of failing to reach Rockall. I wondered who owned this sad lighthouse as we wandered around its derelict rooms and through its little vegetable garden in which potatoes still grew. Vegetable gardens became an essential part of every lighthouse keeper's life, providing not only fresh produce but also a welcome break from the monotony and solitude of life in the light.

For the first time in a week I actually felt hungry. Jamie had decided to make the occasion with a full English roast, and while we ate and drank he launched into his sea shanties, which echoed out across the water. The wind had dropped, the water was calm and I was elevated from my monosyllabic trance into a drunken stupor. Charles had taken the failure in his stride and had already contacted his travel agent in New York by satellite phone to organize onward tickets to the South China Seas, via Stornaway, while Beetle had been thrilled by our little adventure.

Next morning, a bright sun beat down on us, bleaching the still waters with a startlingly dazzling white light. Seagulls wheeled, and seals snorted and coughed as they swam past on their way to catch breakfast. I sat on the deck with my fuggy head and watched a minke whale dipping and diving out in the bay. It was like another world after the previous week.

'I don't believe it,' boomed Jamie's voice from the small navigation post below the helm. He appeared clutching a piece of paper. 'I've just picked up the shipping forecast. A high-pressure ridge is building over Scotland and heading our way, bringing with it fine weather and light winds. With any luck it should give us just enough time to have one last attempt,' he said, beaming.

Charles had already committed himself to a flight to the Far East the following day, but for me it meant one last chance. I was euphoric: I was going to Rockall after all. Third time lucky and all that. The weather had already improved dramatically and if we were to benefit from the lull we had to move fast. We dropped Charles on the nearest island, from whence he could somehow make his way to Harris. Left on an empty Hebridean beach, he was a funny sight dressed in shorts, polo shirt and Nike trainers, with his back-to-front baseball cap and clutching his small trolley suitcase. I wondered what the locals would make of this strange American landing on their shores. He reminded me of the main character in *Local Hero*, the film about a Texan oil prospector who arrives unannounced in a Hebridean village.

As Charles disappeared down the beach we waved goodbye and then for the first time we were able to set *Eda*'s magnificent sails. I swear I heard her sing as we hoisted the mainsail to the sky. The *Eda* danced through the water as the healthy breeze filled her sails as if they were lungs and propelled her across the ocean towards our quarry. Once again we slipped into our watch routines, which were much more pleasant in the absence of the bitter north wind. Once again we passed by a now rather familiar St Kilda. The skies remained clear and for two days we sailed on through waters teeming with dolphins, rafts of puffins and even some unseasonable whales. We hadn't seen another ship for more than a week, but the swell had been reduced from twenty feet to just seven feet. Suddenly it seemed Rockall was in my grasp.

The full moon cast eerie shadows across the deck as we edged our way towards the tiny lump of granite which had become a sparkling diamond in my future crown. The forbidden fruit was tantalizingly near yet still six miles away. Landing at night would be fraught with danger, so we would have to continue tacking up and down until daybreak. At 6.00 a.m., we crossed into 57° 35' 48" N 13° 41' 19" W. A year's journey had culminated in this remote square of the Atlantic map and I caught my first glimpse of Rockall. First it was just a blur on the horizon; then it was a smudge; then a speck; next a dot. Not an island, not a rock: more a pebble. I knew she was small, but this was ridiculous – our mast was taller than she was and we were certainly longer. But there was something ethereal about her. Rockall was beautifully ugly.

Squat and rotund, she had the overall shape of a stumpy pear. The small ledge on her summit looked barely large enough for a single man, let alone for two men to have lived for a month. The small hazardlight had been smashed by the seas that often engulf her with spray five times her height. Fulmars, kittiwakes and guillemots buzzed around the rock while several dozen gannets nested on the guano-streaked summit. A lone seal had wrapped itself in kelp to help anchor it to the rock. It must have been agony for the seal to be unable to scale the sheer cliff. I wondered how it had got there. Perhaps it had entertained the same idea as me.

I had been told that the RAF often used the rock for target practice, flying in from their base in Benbecula on North Uist. A needle in a giant haystack, Rockall was sometimes totally obscured by crashing waves, and pilots often returned home unsuccessful. It seemed incredible that Tom McLean and the Greenpeace invaders had made the tiny ledge their home. It wasn't only narrow and precipitous but totally exposed. You certainly wouldn't want to roll over in your sleep.

For an hour we circled Rockall, the swell breaking against her, engulfing her bulging waist, riding up her body by ten feet before dropping away to reveal her kelp-encrusted petticoat some thirty feet below. I could make out a small foothold from which an ascent looked foolish but possible. Her defences were strong indeed. Tom was at the helm, nudging us a little closer each time we passed. It is not Rockall's visible parts that have caused so much tragedy but her invisible reef: the underwater body of the rock, like an iceberg, is treacherous for shipping. We made several passes, our masts about the same level as her summit, just thirty feet away. I could smell the guano as we edged closer and closer. Jamie was below, monitoring the sonar for the depth. Now was not the time to be shipwrecked.

There was now just twenty feet between us. With each circle Tom had become more and more brazen, but now our speed seemed to have slowed and we were virtually stationary as the *Eda* fought the complex currents. Tom reached for the accelerator and increased the power of the engine, but we remained motionless. He increased the revs, ugly black smoke billowing from the bilges with the effort. The engine screamed and for a split second we appeared to be going backwards, pulled like a magnet to the rock. 'More,' hollered Jamie as we edged closer and closer to disaster. Full throttle, the *Eda* fought to break free of the ocean's grasp, coughing and spluttering before hauling herself through the surf and back out to the safety of the deep water.

Blimey, I thought, my heart pounding with fear. It had been a close shave. We decided to keep our distance. To get any closer we would have to approach her in the dinghy. We had battled for nearly two weeks to reach her, and she was now just thirty feet from me: my island beckoned. Before anyone could say anything, I had packed a knapsack with my new flag, which had been carefully designed by my sister Tamara on some linen bought at London's Portobello market, two flares (for the victory pose), a Havana cigar (from my trip to Cuba), my satellite phone to call my girlfriend (and of course declare her Queen) and a bottle of beer from

North Korea left as a gift from Charles as a gesture of international goodwill. I wondered what the Icelandic and Danish governments would make of this gesture.

I removed a couple of layers to increase my manoeuvrability and donned a lifejacket while Tom and Jamie launched the dinghy into the substantial swell. George, Jamie and I headed out in the dinghy towards my realm. Suddenly Rockall didn't seem so small. The rise and fall of the waves had looked considerable from *Eda*, but from the dinghy they were positively titanic as we disappeared into the troughs. For another hour we circled her, trying to read the swell and work out how best to make an attempt.

The problem was that the waves were erratic, and for every gentle one there were a dozen behemoths that enveloped half the rock. I would have fewer than two seconds in which to grab a handhold in the granite and heave myself six feet up on to Rockall's rotund body and away from the surging swell. At the same time, Jamie would need to motor away to stop the dinghy collapsing backwards into the drop. We circled again, each wave appearing larger than the last, sometimes engulfing half the rock in white water. Rockall was putting up an impressive fight.

'Let's go,' I screamed with as much courage as I could muster. Jamie threw the tiny boat towards Rockall, my hands grabbed at its slippery surface and struggled to get a grip, white water poured over the bow of the boat, the wave dropped and the dinghy raced dangerously backwards. The rock wall slipped from my grasp and I reeled back into the boat. I now realized why past squatters had used a helicopter. The occupation was looking less and less feasible.

'Again, again,' I hollered, as Jamie once again sped towards Rockall. Again my hands grabbed on to its slimy surface, and my right foot hooked into a tiny crevice. I could feel the boat disappearing below me and before I knew it I was alone, clinging tenuously to Rockall. For a split second I wallowed in jubilation, until an enormous wave swept around the corner of the rock, enveloping me in her grasp. A wall of water rose up my body and my life jacket exploded into action, primed by the white water bubbling around my neck. I struggled to cling to Rockall's craggy face, but the pull of the wave was too strong and I felt my grasp torn away by the powerful drag.

Rockall disappeared in a mountain of foam as I tumbled through the waves, gasping for air, and somersaulted through the surf and out into the open ocean beyond. The life jacket had over-inflated, jarring my neck back at an uncomfortable angle. Strangely I felt warm for the first time in two weeks as I bobbed up and down in the waves like a human buoy. I had never imagined I'd be seeing Rockall from this half-submerged perspective. Rockall must have been tantalizing for the seals, not to mention the many men and women who had perished here, unable to land on this seemingly insurmountable rock.

Suddenly I felt a hand on my shoulder and before I realized what had happened, I was hauled aboard the dinghy by the Royal Mail – saved by George, still wearing his postman's uniform. 'I bet that doesn't happen very often on your morning rounds,' I spluttered. George chuckled as I lay on the floor, drenched but safe.

Dwarfed by the mountainous surf, Jamie made for the *Eda*, which had retreated to a safer distance. 'One more time, Jamie,' I implored. 'I have a plan.' It was my last resort. I fished out a small book of yellow Post-it notes from my bag. 'This belongs to Ben,' I scrawled in black ink. If I wasn't going to raise the flag, I was sure as hell going to leave my mark somehow. Jamie raced in once again and, with the aid of some duct tape, I pinned the note to the precipitous wall. Not quite the courageous invasion and bloody coup I had anticipated, but a suitably apt gesture for an island immersed in eccentricity.

I might have touched Rockall, but it was distinctly untouched by me. For now, at least, it remains a realm of the sea, a kingdom governed by the weather, rather than weathered by its governor.

# Epilogue

The sky is streaked crimson as the sun dips into the calm waters of the Bristol Channel. I'm sitting at a small weathered table in a hut atop a cliff overlooking Putsborough beach in North Devon. I can see families packing up their beach balls and picnics, dressing children and shaking sand from their towels. The poet John Betjeman, who loved this region, would have smiled: it is a picture postcard of the British summer seaside idyll. Wild storms and ferocious oceans seem other-timely, otherworldly.

I had been in search of an island of my own, but what I'd found was not so much an island but a way of life and peoples, their lives dominated and structured by their environment. I had seen the best and the worst, the highs and the lows of island life, and, above all, its stark reality. Like miniature portraits, the islands, isles, reefs, skerries and forts had offered a glimpse of lives past and present, harsh and ideal, beautiful and ugly. I had met not my perfect island match but lands of contradiction that lured me to them as honey draws a bear.

They may languish on the periphery of our national consciousness, but these offshore lands are an integral part of our collective identity, living pieces of our historical jigsaw. I had been surprised by their unique individuality, often encapsulated in and indeed shaped by their rich histories.

How quickly times shift offshore, like the waters and sands around the islands, waxing and waning with each breath of wind and surge of tide! Much has changed in the few months since my visits.

The Puff Inn on St Kilda closed because of the threat of terrorism. Patrick and Gwyneth Murphy left Bardsey. The success story of Eigg precipitated what the press affectionately termed a 'Mugabe-style land grab' in the form of community buyouts of Harris, Gigha and South Uist. Muck advertised for another new family. The *Patricia* got a new helicopter. The NatWest Island Games moved to Shetland, where the Vikings behaved themselves. The Sealand international football team were beaten by Denmark, and Warner Bros have begun production of the film about Sealand, with Russell Crowe set to play Prince Michael. Meanwhile Madonna was dropped from hubby Guy Ritchie's film *Revolver*, which received the worst reviews of recent times. Eilean Mullagrach in the Summer Isles was put back on the market and I missed out once again. Danny Wallace set up his own country – in his flat in Bow, East London; and Charlie Connelly gave up his attempt to invade Rockall.

Meanwhile Rockall was assaulted and successfully claimed by the *Rockall Times*. And I remain island-less.

# Bibliography

Bella Bathurst, *The Lighthouse Stevensons*, Flamingo, 1999

Charlie Connelly, *Attention All Shipping*, Little Brown, 2004

Norman Davies, *The Isles*, Papermac, 2000

George Drower, *Heligoland*, Sutton Publishing, 2002

James Fisher, *Rockall*, The Country Book Club, 1957

Hamish Haswell-Smith, *The Scottish Islands*, Canongate, 1999

Peter Hill, *Stargazing*, Canongate, 2004

Charles Maclean, *St Kilda*, Canongate, 2001

Adam Nicolson, *Seamanship*, HarperCollins, 2004

James and Deborah Penrith, *Orkney and Shetland*, Vacation Work, 2002

John Prebble, *The Highland Clearances*, Penguin, 1969

Jonathan Raban, *Coasting*, Collins Harvill, 1986

George Rosie, *Curious Scotland*, Granta Books, 2004

W. G. Sebald, *The Rings of Saturn*, Vintage, 1998

Erwin S. Straus, *How to Start Your Own Country*, Breakout Productions, 1999

Leslie Thomas, *My World of Islands*, Mandarin, 1995

Peter Unwin, *The Narrow Sea*, Review, 2004